water pollution technology

water pollution technology

JOHN A. BLACK

Suffolk County Community College
Selden, New York

RESTON PUBLISHING COMPANY, INC.
Reston, Virginia

A Prentice-Hall Company

Library of Congress Cataloging in Publication Data

Black, John A.
Water pollution technology.

Includes bibliographies and index.
1. Water–Pollution. 2. Water–Pollution–Measurement. 3. Sewage–Purification. I. Title.
TD420.B58 628 76–27740
ISBN 0–87909–875–9

Reston Publishing Company, Inc.
A Prentice-Hall Company
Reston, Virginia 22090

10 9 8 7 6 5 4 3 2 1

Printed in the United States of America

To

GAIL

who has carried much of
the burden in the preparation
of this book

contents

preface

In this text I have attempted to bring together the total spectrum of information on marine, surface, and groundwater systems; the sources and consequences of pollution and possible solutions; as well as applications and the analytical methods commonly used in water and wastewater monitoring.

Since all aquatic systems operate by the same or similar chemical, physical, and biological principles, much of the information is interrelated and is as pertinent to deep oceans as it is to freshwater bogs. Hence many principles, methods, and applications are discussed extensively in earlier chapters and merely referred to in later sections. Also included are detailed sections explaining instrument use and standardization, as well as the pertinent analytical methods commonly used in pollution monitoring.

The reader will, therefore, obtain general background information as well as explicit analytical techniques and applications used in monitoring environmental pollutants, together with an understanding of the consequences of releasing these materials into waterways.

It is the author's opinion that this text may be used at a variety of instructional levels. For example, the general background information can be used in introductory environmental curricula; and, when combined with the applications and analytical techniques, the material is suitable for more advanced students.

I would like to express my appreciation to the many individuals and organizations for assistance during the preparation of this book. Special thanks to Fred Drewes, Ken Ettlinger, Jim Schramel, and Sue Stehlin of Suffolk County Community College, and to Ron Rozsa, a graduate student at the University of Connecticut for commenting on various portions of the preliminary manuscript and for helping with illustrations and photographs. I would also like to acknowledge the contributions of Violet Schirone of Suffolk County Community College for the material on microbiology, and Richard Horner of Northampton Area Community College for suggested additions to the sections dealing with practical applications. Jim Tripp of The Environmental Defense Fund provided help with the chapter on the legal aspects of water pollution control, and Pat Dugan of The Suffolk County Water Authority aided with the appendixes, which deal with analytical methods. Thanks are due also to Karen Chanley for the fast, accurate typing of the many drafts of the manuscript, and to Larry Benincasa of Reston Publishing Company, Inc.

John A. Black

CHAPTER 1

chemical and physical properties of water

Prior to discussing natural, aquatic, and marine systems, it is necessary to discuss the chemical and physical properties of the constituents that comprise these systems. Although the waters of the world contain many, if not all, of the chemical elements known to man, these elements have many commonly shared characteristics. Thus a consideration of the chemical and physical characteristics of these elements provides a convenient starting point for the study of water resources.

general physical–chemical principles

All matter is composed of electrically neutral atoms, and all atoms are composed of smaller components termed subatomic particles. Although physicists have discovered over 30 subatomic particles, only 3 need be considered here: the proton, the neutron, and the electron.

The *proton* is a positively charged particle located in the center of the atom in a region termed the *nucleus*. The proton is a relatively stable particle and has an assigned relative weight of 1. The *neutron* is also located in the nucleus, is neutrally charged but unstable, and also has an assigned relative weight of 1. *Electrons* are negatively charged, orbit the nucleus in a series of electron clouds, are relatively stable, and have an assigned relative weight of 1/1840th the weight of a proton or neutron. Since all atoms are electrically neutral, they must have the same number of electrons as protons in order to maintain electrical neutrality. The sum of the weights of these subatomic particles gives each atom its characteristic weight, which is termed the *gram atomic weight* of that atom. Since it is impossible to weigh and/or work with individual atoms, chemists generally work with larger quantities, such as grams. To maintain a consistent relationship between atoms they use quantities termed moles. A *mole* of any substance is merely the gram atomic weight of that substance expressed in grams.

The number of protons contained in the nucleus of a particular atom, as well as the gram atomic weight of the atom, are conveniently determined from the *Periodic Chart* (Fig. 1-1). Note that each atom on the chart is represented by a symbol and that numbers appear both above and below the symbol for each atom. The number directly below the symbol represents the atomic weight of that element, while the number above the symbol gives the total number of protons contained within the nucleus. Since all atoms are electrically neutral, there must be an identical number of electrons circling about the nucleus in one or more electron clouds. For example, hydrogen is listed on the periodic chart as $\underset{1.008}{\overset{1}{\mathrm{H}}}$. The number 1 indicates that the hydrogen atom has one proton in the nucleus and, hence, must have an electron cloud consisting of one electron. The lower number (1.008) is the gram atomic weight of the hydrogen atom. Thus if a chemist weighed out 1.008 grams (g) of hydrogen, 1 mole of hydrogen would be obtained.

Table 1-1 summarizes the characteristics of the subatomic particles. It is to be noted that the only basic difference among the various atoms is their size and weight, which is governed by the number of protons, electrons, and neutrons that comprise any given atom.

TABLE 1-1

characteristics of subatomic particles

Particle	*Symbol*	*Charge*	*Relative Weight*	*Stability*
Proton	+	Positive	1	Stable
Neutron	0	Neutral	1	Unstable
Electron	e^-	Negative	1/1840	Stable

When two or more atoms react, they do so by completely or partially transferring electrons and thus forming molecules. In general, there are three basic types of chemical reactions that occur. The first type of reaction involves the complete loss of electrons by one atom or atoms and the complete gain of electrons by another atom or atoms within the molecule. In reactions of this type, the atom losing electrons (losing units of negative charge) becomes positive, because it has an excess of protons. The atom gaining electrons (gaining units of negative charge) becomes negative, since it has an excess of electrons. The molecules formed in these reactions are termed *ionic compounds*. A

Period or principal quantum number	I	II											III	IV	V	VI	VII	VIII
1	1 H 1.0080																	2 He 4.0026
2	3 Li 6.939	4 Be 9.012											5 B 10.81	6 C 12.011	7 N 14.09	8 O 15.999	9 F 18.998	10 Ne 20.18
3	11 Na 22.99	12 Mg 24.31											13 Al 26.98	14 Si 28.09	15 P 30.98	16 S 32.06	17 Cl 35.46	18 Ar 39.95
4	19 K 39.10	20 Ca 40.08	21 Sc 44.96	22 Ti 47.90	23 V 50.95	24 Cr 52.01	25 Mn 54.94	26 Fe 55.85	27 Co 58.93	28 Ni 58.71	29 Cu 63.54	30 Zn 65.38	31 Ga 69.72	32 Ge 72.59	33 As 74.91	34 Se 78.96	35 Br 79.92	36 Kr 83.80
5	37 Rb 85.48	38 Sr 87.63	39 Y 88.92	40 Zr 91.22	41 Nb 92.91	42 Mo 95.95	43 Tc	44 Ru 101.1	45 Rh 102.91	46 Pd 106.4	47 Ag 107.88	48 Cd 112.41	49 In 114.82	50 Sn 118.70	51 Sb 121.76	52 Te 127.61	53 I 126.91	54 Xe 131.30
6	55 Cs 132.91	56 Ba 137.36	57 La* 138.92	72 Hf 178.50	73 Ta 180.95	74 W 183.86	75 Re 186.22	76 Os 190.2	77 Ir 192.2	78 Pt 195.09	79 Au 197.0	80 Hg 200.61	81 Tl 204.39	82 Pb 207.21	83 Bi 209.00	84 Po	85 At	86 Rn
7	87 Fr	88 Ra 226.05	89 Ac**															

Legend ... Atomic number / Element / Atomic mass amu

*Lanthenum series
**Actinum series

58 Ce 140.13	59 Pr 140.92	60 Nd 144.27	61 Pm	62 Sm 150.35	63 Eu 152.0	64 Gd 157.76	65 Tb 158.93	66 Dy 162.51	67 Ho 164.94	68 Er 167.27	69 Tm 168.94	70 Yb 173.04	71 Lu 174.98
90 Th 232.04	91 Pa	92 U 238.07	93 Np	94 Pu	95 Am	96 Cm	97 Bk	98 Cf	99 Es	100 Fm	101 Md	102 No	103 Lw

FIGURE 1–1

Periodic chart.

typical ionic compound is formed as the result of the reaction between sodium (Na) and chlorine (Cl) in which the sodium chloride molecule (NaCl) is formed. In this reaction electrons are lost by the sodium and gained by the chlorine. This transfer of electrons gives sodium a positive charge and chlorine a negative charge. The resultant molecule may be represented as $Na^{+} \rightarrow Cl^{-}$, with the arrow serving to indicate the direction of electron movement.

The second type of reaction involves the complete and equal sharing of electrons. Reactions of this type occur only between identical atoms. In these reactions the electrons are equally shared, so no charge develops on any of the atoms comprising the molecule. These molecules are termed *nonpolar, covalent molecules.* A typical molecule of this type is hydrogen gas, (H_2), in which two hydrogen atoms come together and equally share electrons. Since the electrons are shared by identical atoms, there is no tendency for one atom to attract electrons more or less strongly than the other, and no charge develops. This molecule may be pictured as $H \longleftrightarrow H$.

The third type of reaction involves the unequal sharing of electrons between the atoms that comprise the molecule. The atom or atoms that partially lose their electrons assume a partial positive charge, whereas the atoms that partially gain the electrons assume a partial negative charge. Molecules formed in this manner are termed *polar, covalent molecules.* A typical polar covalent molecule is water (H_2O), in which the two hydrogen atoms partially lose their electrons to an oxygen atom. Thus the hydrogen atoms assume a partial (α) positive charge, and the oxygen assumes a partial negative charge. The molecule may be represented as $H^{\alpha+} \rightarrow O^{\alpha-} \leftarrow H^{\alpha+}$.

chemical and physical properties of water

In the case of the water molecule it has been found that, in addition to its being polar, the atoms are arranged in a *bent-chain configuration.* The water molecule will, henceforth, be pictured as $_{+}\diagup^{-}\diagdown_{+}$, where the positive regions (actually partially positive) are the hydrogen atoms, and the negative regions (actually partially negative) are the oxygen atoms.

Because of its polarity, water molecules tend to arrange themselves in a definite, ordered manner. This is due to the fact that the positive region of one molecule attracts the negative region of an adjacent molecule, forming a relatively weak intermolecular attractive force termed a *hydrogen bond.* The positive ends of this molecule will then form hydrogen bonds with the negative region of still another molecule, and so on. The end result can be visualized as a long series of water molecules arranged so that the positive ends of various water molecules are aligned with the negative ends of adjacent molecules, and vice versa. This is schematically illustrated in Fig. 1-2A.

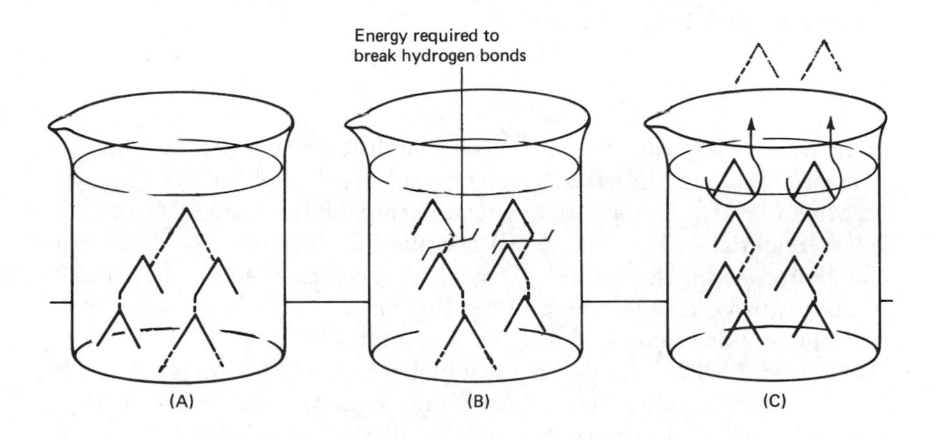

FIGURE 1–2

(A) Schematic representation of a group of hydrogen-bonded water molecules. **(B)** Sufficient energy must initially be added to break the hydrogen bonds. **(C)** Additional energy must then be added to allow the molecules to break through the surface and enter the vapor state.

hydrogen bonding

Hydrogen bonds are of the utmost importance in giving water its unique chemical and physical properties. This is readily illustrated by considering the phenomenon that occurs when a given liquid is brought to its boiling point. In order to boil a liquid, energy must be added. The energy, in the form of heat, is absorbed by the liquid and serves to increase the molecular motion of the molecules comprising the liquid. Once the molecules are moving at a sufficient rate, they are able to break through the surface of the liquid and enter the vapor state. At this point the liquid is said to boil. In the case of a polar liquid, such as water, sufficient energy must not only be added to achieve the stage of high molecular motion, but, prior to reaching this stage, initial energy must be put into the system to first break the hydrogen bonds that hinder free molecular motion of the molecules within the liquid (Fig. 1-2B and C). For example, if it requires four units of energy to place the molecules in the vapor state, an additional initial amount of energy must be added to break the hydrogen bonds. Consequently, that liquid would have a high boiling point (a higher temperature would be required to make the liquid boil). It has been calculated that if water were a nonpolar liquid, it would have a boiling point of approximately −80°C rather than its normal boiling point of 100°C. In other words, if water were nonpolar, it would be a vapor at normal earth temperatures.

density

The relationship of hydrogen bonding to temperature is extremely important in natural systems and can be related by the concept of density. Density is the relationship of the mass (M) of any given material to the volume (V) occupied by that material. This can be expressed by the formula $D = M/V$. According to this equation, density can be changed by altering the weight (mass) or volume of a substance. For example, if a liquid has a mass of 10 g and occupies a volume of 10 liters, its density would be 1 ($D = M/V$ or $D = 10/10 = 1$). If, on the other hand, the volume remains constant at 10 liters and the mass is decreased to 5 g, the density would decrease to 0.5 ($D = 5/10 = 0.5$). Thus the density of a particular substance will increase if the mass is increased and the volume kept constant, or if the volume is decreased while the mass remains constant (or increases). Conversely, the density will decrease if the mass is decreased and the volume held constant (or increased), of if the volume is increased and the mass is held constant (or decreased). Table 1-2 summarizes the possibilities.

TABLE 1-2

means of changing density

Assuming that the initial conditions are as follows:
$M = 2$ g; $V = 10$ liters; $D = 2/10 = 0.2$

Factors increasing D:

1. If M is increased to 4 g with V remaining constant (10 liters)	$D = 4/10 = 0.4$; density is increased to 0.4
2. If M is constant (2 g) with V decreased to 5 liters	$D = 2/5 = 0.4$; density is increased to 0.4

Factors decreasing D:

1. If M is constant (2 g) with V decreased to 20 liters	$D = 2/20 = 0.1$; density is decreased to 0.1
2. If M is decreased to 1 g with V constant (10 liters)	$D = 1/10 = 0.1$; density is decreased to 0.1

temperature

As noted previously, hydrogen bonding plays an important role in determining the density of natural marine and freshwater systems. This role is established through the effects of temperature on the formation and amount of hydrogen bonds formed within a given system. The degree of hydrogen bonding will affect the volume of a given system, and this, in turn, will affect the density. As discussed previously, energy, in the form of heat, must be added to water to not only allow the molecules to attain sufficient speed to enter the vapor state, but (initially) sufficient heat must also be added to break the hydrogen bonds which prevent unrestricted movement of the individual water molecules. Thus by raising the temperature of the liquid water, hydrogen bonds are broken and the molecules can move about more freely. In other words, molecular motion is increased. By increasing molecular motion, the distance between the individual molecules is increased. This increases the volume occupied by the liquid and decreases the density. In general, the higher the temperature of any given system, the lower the density. The reverse occurs when a liquid is cooled. As the temperature is decreased, the molecules move more slowly. Since molecular motion is decreased, hydrogen bonds can be easily formed between adjacent, slow-moving molecules. Since increased hydrogen bonding restricts the molecular motion even further, less volume is occupied by the liquid. As the volume decreases, the density must increase.

The effects of temperature on hydrogen bonding, and thus on density, can be illustrated by considering the phenomenon that occurs when water is cooled from a very high temperature to 0°C. Assume a hypothetical case where 10 g of water is present in a sealed 1-liter container at a temperature of 105°C (this is above the boiling point and, therefore, all the water can be considered to be in its vapor state). With all the water in the vapor state, the water is considered to be effectively occupying the entire container, since the pressure exerted by the water is identical on all the walls of the container. This pressure is exerted by the water molecules, moving independently and colliding with the walls of the container. At such a high temperature it can be assumed that there are no hydrogen bonds present in the system. If the container is cooled to 90°C, for example, the water molecules would begin to slow down. As their speed decreases should one water molecule, at this lower temperature, encounter another molecule, there is a good probability that they would tend to form hydrogen bonds and become "heavier" than the medium in which they are floating. At this point the water will condense (go into the liquid state). For purposes of illustration, assume that at 90°C essentially all the molecules have formed hydrogen bonds to some degree, all the water is in the liquid state, and the molecules, at this temperature, now occupy only 75% of the container (the volume has decreased to 0.75 liter). Since the container is sealed, no molecules escape or enter; and the mass has therefore remained constant while the volume has decreased, owing to the condensation of the water vapor, and the density has increased (Fig. 1-3).

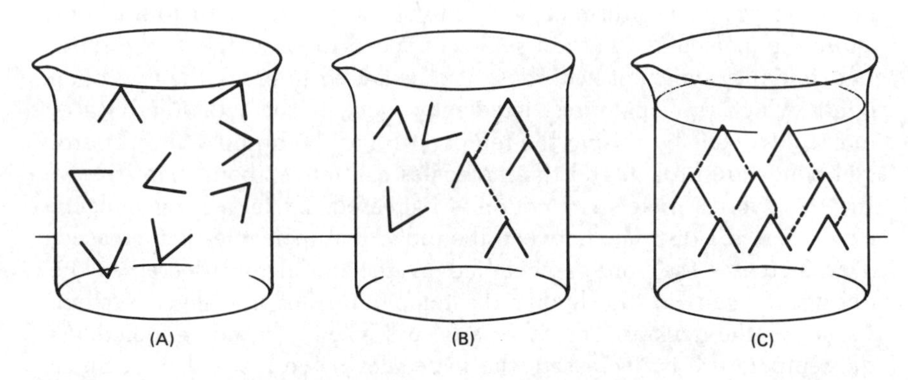

FIGURE 1–3

(A) Schematic representation of water at 105°C. Note that all of the molecules are moving independently and that hydrogen bonds are absent. At this point the water is effectively occupying the entire container. **(B)** As the water is cooled below the boiling point, hydrogen bond formation begins. The water will begin to condense and enter the liquid state. **(C)** At this point condensation is essentially complete. The water is in the liquid state, and the volume occupied by the water has decreased.

If the temperature is lowered to 50°C there will be even less energy to move the molecules about. Consequently, more molecules will be moving at a lower speed (in the liquid state) and more hydrogen bonds will be able to form. Since hydrogen bonding is increased further, molecular motion is decreased, and this will result in a decreased volume. As the volume decreases, the density will increase further, and the liquid will become "heavier." In reality, however, the mass is not increasing; rather, the same number of molecules are, at this stage, packed into smaller and smaller space. Thus the volume is decreasing and the density is increasing.

This phenomenon of water increasing in density as the temperature decreases will continue until the water reaches a temperature of 4°C. At this temperature water is at its maximum density with maximum hydrogen-bond formation and with minimum volume occupied by the water molecules. As water is cooled below 4°C, the density actually decreases, and, therefore, this water becomes progressively "lighter" (less dense) and floats on top of the warmer water. When water reaches the freezing point, it is at its lowest density, and ice, therefore, floats above the warmer, denser water in the hypothetical container.

To explain this atypical behavior, it is necessary to examine the molecular structure of the actual ice crystal (Fig. 1-4). Water at the

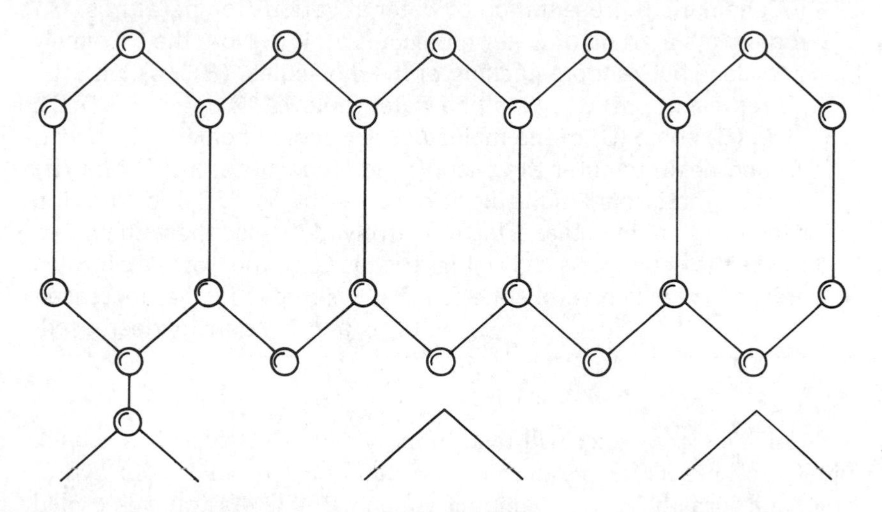

FIGURE 1–4

Schematic representation of several ice crystals. Each crystal consists of six water molecules (represented by the circles) connected by hydrogen bonds (represented by the solid lines).

freezing point assumes a rigid, hexagonal crystalline structure. As the water molecules enter the solid state (ice), they tend to spread out and rearrange themselves into this definite, ordered crystalline structure. Because of the bond angles there is a large amount of open space between the adjacent water molecules that participate in the ice crystals.

Prior to entering the true crystalline state, however, the liquid water molecules tend to rearrange themselves (at temperatures below 4°C but above 0°C). These initial rearrangements are termed the pseudocrystalline states of water. In the course of these preparatory rearrangements, the water molecules begin to spread apart. This increases volume somewhat and partially decreases the density. This process is schematically summarized in Fig. 1-5.

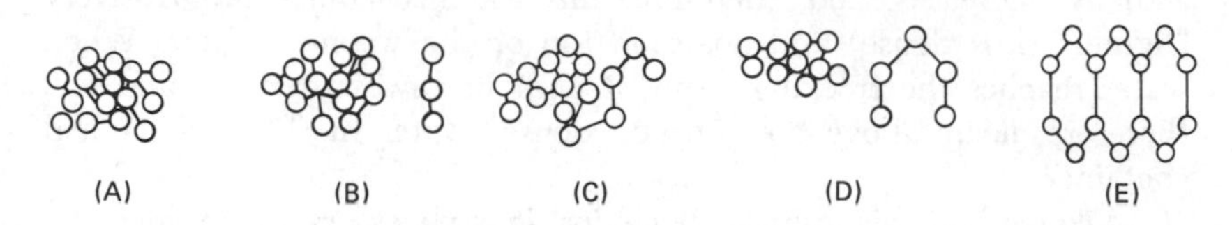

FIGURE 1–5

Schematic representation of water at various temperatures. (**A**) represents a group of water molecules at 4°C. Note the extremely close but random packing of the molecules. (**B**), (**C**), and (**D**) represent pseudocrystalline water (below 4°C but above 0°C). In (**B**) some (3) of the molecules are cooled below 4°C, spread out, and begin to enter the pseudocrystalline state. In (**C**) and (**D**) additional molecules are cooled below 4°C and enter the pseudocrystalline state. This effectively increases the volume and lowers the density. As the water reaches 0°C, the water molecules enter the true ice crystal state (**E**). The volume is further increased and the density decreased.

In summary, water will tend to increase in density as it is cooled, until it reaches its temperature of maximum density, maximum hydrogen bond formation, and minimum volume at 4°C. As water is cooled below 4°C, it goes through its pseudocrystalline states, during which the molecules rearrange, spread out, and increase in volume prior to entering the open crystalline ice structure. Therefore, ice has a lower density than liquid water at any temperature.

As water above 4°C is cooled, the density increase that is noted is due to reduced thermal agitation. This causes the water to move more slowly, thereby increasing the possibility of hydrogen-bond formation and the decrease in volume. As water is cooled below 4°C, the reduction in volume does not compensate for the increased numbers of water molecules entering their pseudo and truly crystalline forms. It is this increased number of these molecules that is responsible for the increased water volume occupied by water molecules at temperatures below 4°C. Density effects on natural systems are discussed in Chapters 4 and 6.

water: the universal solvent

In addition to its influence on density, the polarity of the water molecule is responsible for it being termed the *universal solvent.* If an ionic compound such as sodium chloride (NaCl) is placed in contact with water it readily dissolves. It will be recalled that electrons are transferred from the Na atom to the Cl atom in the formation of the molecule. The sodium has, therefore, assumed a positive charge since it has lost an electron to the Cl, which has developed a negative charge in the process. Thus the molecule may be drawn as $Na^{+} \rightarrow Cl^{-}$. When a polar liquid such as water comes in contact with ionic molecules or polar compounds (Fig. 1-6), the water will orient itself around these

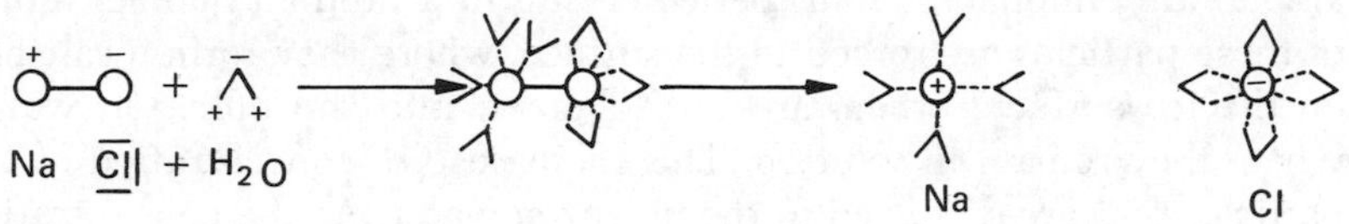

FIGURE 1–6

When polar water molecules encounter an ionic compound such as NaCl, the water orients about the NaCl so that the negative portion (O) of the water aligns with the positive region (Na) of the ionic compound, and the positive portion (H) aligns with the negative region (Cl). Similar orientations occur when water encounters polar molecules.

molecules so that the positive ends of the water align with the negative regions of the molecule, and water's negative regions align with the molecule's positive portions.

This intermolecular attraction between the water (the solvent) and NaCl (the solute) tends to break the chemical bond between the sodium and the chlorine, and the molecule separates into its component parts. This type of separation is termed *dissociation.* It is to be noted that, when the bond breaks, the electron that sodium lost in the initial NaCl formation remains with the chlorine. Consequently, sodium goes into solution as a positively charged particle (ion) and chlorine as a negatively charged particle termed the *chloride ion.*

In determining the tendency of a given solute to dissolve in a given solvent, it is useful to make the generalization that "like dissolves like." In other words, a highly polar solvent such as water will tend to readily dissolve ionic or polar solutes, whereas nonpolar or only slightly polar solutes tend to be insoluble.

When liquid water with any appreciable amount of dissolved material is cooled, it is found that the point of maximum density is not at 4°C but is, rather, at the point of lowest temperature, regardless of the temperature that the solution attains below 0°C. This is due to the tendency of the ice crystal to exclude foreign (nonwater molecules) dissolved particles from its hexagonal crystalline structure.

As noted previously, ice has a considerable amount of space between the water molecules that comprise the individual ice crystals. Regardless of the open structure of the ice crystal, foreign particles are not able to fit within the interstices between adjacent water molecules. As ice forms, these dissolved particles (Cl^- and Na^+, for example) are eliminated from the ice crystal in a random manner—some of these particles are forced to the surface, where they form a salt film on the ice surface; others are forced down into the unfrozen water, where they remain in solution. This increases the concentration of the particles that are dissolved in the unfrozen water. As the concentration of the solute increases, the mass of the solution also increases. As ice formation continues, the volume of the liquid water must decrease, since water is leaving the liquid state and entering the solid state (forming ice). The volume of this solution is, therefore, decreased as the mass is increased. This results in an increased density of the solution.

To summarize, it can be considered that in solute–solvent systems, the colder the water, the greater amount of ice that is formed. As ice formation increases, a greater number of dissolved particles will be excluded from the ice crystal and forced into the liquid water, while the volume of the liquid is reduced. This effectively increases the density of the solution. In these systems, therefore, the point of maximum density is not at 4°C; the density is governed by the dissolved particle concentration of the remaining unfrozen liquid.

This phenomenon of the *freezing out* of salts also plays an important role in decreasing the freezing point of solutions. It is found that when pure water is cooled, it freezes at 0°C. However, if any given substance such as sugar is dissolved in the water and the solution is then frozen, it will have a freezing point below 0°C. When a specific number of particles are dissolved in 1 kilogram (kg) of water or any liquid, it will lower the freezing point of the solution by a quantitative, predictable amount. This reduction is termed the *Molal Freezing-Point Depression.*

As noted previously, there is a common term, the mole, that

gross refers to 144 units, the mole refers to 6.02×10^{23} particles (atoms, molecules, etc.). Whereas a dozen refers to 12 units and a gross refers to 144 units, the mole refers to 6.02×10^{23} particles and is termed *Avogadro's Number,* in honor of Amedo Avogadro, the chemist who first hypothesized its existence. Since the mole refers to the gram atomic weight of any element expressed in grams, it can be determined from the Periodic Chart that hydrogen has a gram atomic weight of 1.008. Therefore, 1 mole of H would weigh 1.008 g and would contain 6.02×10^{23} hydrogen atoms. One mole of chlorine (gram atomic weight = 35.453) would weigh 35.453 g and contain 6.02×10^{23} chlorine atoms. In each case it is to be noted that although the weights change, the number of atoms contained in a mole will remain the same (just as a dozen eggs will weigh more than a dozen feathers, yet each dozen contains 12 units).

When one considers moles of molecules the situation is similar. The NaCl molecule consists of one Na atom chemically bonded to one Cl atom. One mole of NaCl would be composed of 6.02×10^{23} Na atoms with each Na bonded to a Cl atom. Thus there must be an identical number of Cl atoms (1 mole, or 6.02×10^{23}). Consequently, in 1 mole of NaCl there is 1 mole of molecules. Each molecule consists of one Na and one Cl atom; therefore, in 1 mole of NaCl molecules there are two moles of atoms (1 mole of Na atoms and 1 mole of Cl atoms).

In 1 mole of $CaCl_2$ ($Cl^- \leftarrow Ca^+ \rightarrow Cl^-$) there would be 6.02×10^{23} $CaCl_2$ molecules. Each molecule would consist of 2 Cl atoms and 1 Ca atom. One mole of $CaCl_2$ would contain 3 moles of atoms (1 mole of Ca atoms and 2 moles of Cl atoms) and would weigh 110.986 g (the sum of the gram atomic weight of each atom expressed in grams). One mole of $Mg(OH)_2$ ($H^+ \rightarrow O^- \leftarrow Mg^+ \rightarrow O^- \leftarrow H^+$) would contain 5 moles of atoms, since each $Mg(OH)_2$ molecule contains five atoms, and would weigh 58.324 g (the sum of the gram atomic weight of each atom expressed in grams).

As mentioned previously, when a given quantity of particles is dissolved in 1 kg of water, the freezing point of the solution will be lowered by a quantitative amount. When 1 mole of particles is dissolved in 1 kg of water, the freezing point will be decreased to −1.86°C; in other words, the freezing point will be lowered below 0°C. It makes no difference whether it is 1 mole of Cl, 1 mole of K, or 1 mole of Cu that is dissolved. In each case, with 1 mole of solute dissolved in 1 g of water, the freezing point will be decreased by 1.86°C. This is known as the *Molal Freezing-Point Depression Constant* of water. When predicting the freezing points of solutions that contain dissolved molecules, it is necessary to consider the process of dissociation (p. 11). It will be recalled that when NaCl dissolves, it dissociates into Na^{+} and Cl^{-} ions. Thus if 1 mole (58.442 g) of NaCl were dissolved in 1 kg of water, the resulting solution would freeze at −3.72°C. This is due to the fact that when NaCl is dissolved in a polar solvent, it dissociates into its component parts. Therefore, 1 mole of NaCl will dissociate into 1 mole of Na^{+} ions and 1 mole of Cl^{-} ions. Consequently, the resultant solution will contain 2 moles of particles, and the freezing-point depression will be doubled (2 × 1.86°C). If a polar molecule such as $CaCl_2$ is dissolved, each molecule will dissociate into one Ca^{+2} ion and two Cl^{-} ions. Therefore, 1 mole of $CaCl_2$ molecules dissolved in 1 kg of water will depress the freezing point to −5.58°C, since there are 3 moles of ions in the resultant solution.

Freezing-point depressions in solutions occur because the solute actually dilutes the solvent. The physical presence of the solute forces the solvent molecules farther apart to provide space for the solute. In order to freeze, the solvent molecules must come into close proximity with each other to allow the hydrogen bonds, needed to produce the crystalline structure, to form. In order for this to occur, the solvent molecules must be made to move at a lesser rate than normal. Consequently, additional heat must be removed from the system to bring about the desired decrease in molecular motion. In other words, the temperature must be lowered below the normal freezing point.

As a solution freezes, the unfrozen liquid portion not only increases in density due to the freezing out of the salts, but its freezing point is also depressed. Consequently, the colder the water gets, the denser it becomes, and the harder it is to freeze the remaining liquid. Eventually, in natural marine systems, a point is reached where it is impossible to attain the very low temperatures necessary to freeze the extremely "salty" water. The water, at this point, will remain as a very dense, highly salty liquid water mass.

In addition to the above mentioned properties of water, the following properties are also related to hydrogen bonding and are of importance in the functioning of marine and freshwater systems.

solubility of gases

Boiling water (or merely increasing the temperature) will tend to lower the concentration of dissolved gases. It is to be recalled that when pure water is heated, more and more water molecules gain sufficient energy to allow them to escape from the liquid state. Similarly, if a beaker of water containing a dissolved gas is heated, the solute (gas) will gain energy and eventually attain a sufficient velocity to enable it to overcome the attractive forces and escape from the surface of the liquid, thus decreasing the concentration of the dissolved gas in solution. The reverse will occur if the water is cooled. In general, a given volume of cold water will contain a greater concentration of dissolved gas than an equal volume of warm water. Also, nonpolar gases such as O_2 are less soluble in water than polar gases such as O=C=O (CO_2).

specific heat

Of all the elements and compounds known to man, only lithium, ammonia, and liquid hydrogen are capable of storing a greater quantity of heat with a smaller temperature rise. Because of water's "heat-storing capacity" it is said to have a high *specific heat.* Since very large quantities of heat are involved in aquatic temperature changes, natural systems (both marine and aquatic) warm up slowly in the spring and cool slowly in the fall, thus preventing wide seasonal temperature fluctuations and thereby moderating terrestrial temperatures.

latent heat

Water has the highest heat of fusion and heat of evaporation (latent heat) of any naturally occurring liquid. Heat of fusion refers to the large quantities of heat that must be removed to form ice and, conversely, the large quantities necessary to melt the ice. It is calculated that it takes 80 calories of heat to freeze 1 g of water—that is the equivalent to the number of calories required to raise the same quantity of liquid water at 0 to 80°C. Consequently, a body of water at 0°C can lose sufficient heat to warm a large amount of cold air with only minimal ice formation.

In evaporation (heat of evaporation), 536 calories is needed to convert 1 g of liquid water to vapor at a temperature of 100°C. This is equivalent to the amount of heat required to raise 536 g of water 1°C and is primarily due to the polar nature of water. The heat absorbed in evaporation is released in condensation and plays a major role in worldwide temperature changes.

SUGGESTED READINGS

Seager, Spencer L., and H. Stephen Stoker. 1973. *Chemistry: A Science for Today*. Glenview, Ill.: Scott, Foresman and Company.

Sienko, Michael J., and Robert A. Plane. 1976. *Chemistry*, 5th ed. New York: McGraw-Hill Book Company.

QUESTIONS

1 / Explain the difference between ionic and covalent molecules.

2 / Explain how polar covalent and nonpolar covalent molecules differ.

3 / Explain why the polar nature of water gives the water molecule its unique chemical and physical properties.

4 / Explain how hydrogen bonds differ from true chemical bonds.

5 / Explain the relationship of hydrogen bonds, temperature, and molecular motion.

6 / Describe what occurs as water temperature is lowered from 100°C to 0°C.

7 / Explain why ice floats.

8 / What is the difference between the solvent and the solute?

9 / Explain how the addition of solute lowers the freezing point.

10 / Why is water termed the universal solvent?

CHAPTER 2

microbial systems

In all natural systems certain materials are taken in by plants, converted to complex molecules, and incorporated into plant tissues. These materials then pass into various animals as a result of the animals grazing on the plants. When the plants die and the animals excrete waste materials or die, these materials are reconverted by a large group of organisms, termed *microorganisms,* into a form where they can again be used by plants. Thus materials tend to cycle from plant to animal to microorganism and back to plants in what is termed a *food chain.*

Microorganisms, also known as *protists,* are vital to the function of food chains since it is by their activity that plant and animal residues are converted into the proper form for plant use. Thus all life ultimately depends upon the functioning of microbial systems.

Unlike the larger organisms which exhibit differentiation and specialization of cells into tissues, organs, and so on, the microorganisms are single-celled and perform all their necessary life functions within their single cell. Microorganisms include bacteria, blue-green algae, fungi (yeasts and molds), green and brown algae, the protozoans, and the viruses. The viruses, unlike other protists, are noncellular and exist as naked protein. Except for the viruses and the bacteria *Chalmydia* and *Rickettsia,* all protists are capable of independent life.

microbial relationships

Microorganisms rarely occur in natural systems as pure cultures consisting of only one type or single population, but occur as mixed populations or communities containing several different types. Microbial communities are constantly changing in response to both environmental factors and the interactions of other organisms within the microbial, plant, and animal community.

The interaction among organisms of different species is termed *symbiosis.* The symbiotic relationship may result in one of three direct effects that may be exerted on the participating organisms: commensalism, mutualism, or parasitism. *Commensalism* is an example of a symbiotic relationship in which one organism benefits from the association while the other organism is unaffected. *Mutualism* is a relationship in which each organism involved derives benefit. In *parasitism* one organism derives benefit while the other, termed the host, is harmed. Parasitism is distinct from predation in that a predator ingests its prey while a parasite insidiously removes materials from the host and in so doing derives nutrients in a manner that gradually harms or weakens the host.

There is an indirect relationship, known as *saphrotism,* which is important in coastal marine systems. Saphrophytes are involved in the breakdown of various materials, such as marsh grass, into smaller particles termed *detritus.* The detritus is then utilized as a food source by the animals present in these systems. Thus, in the saphrophyte chain, material cycles from plants to microorganisms to animals, rather than from plant to animal to microorganism as previously discussed. The saphrophyte food chain is discussed in Chapter 6.

types of microbial decomposition

In natural systems microorganisms accomplish the decomposition of plant and animal residues by three major mechanisms: aerobic, anaerobic, and facultative decomposition. The bacteria active in performing these processes are termed aerobic, anaerobic, and facultative bacteria.

Aerobic decomposition requires the presence of oxygen, and this oxygen is converted to water in the process. In *anaerobic decomposition* the presence of oxygen is not required; other materials, such as nitrite (NO_2) and nitrate (NO_3), are utilized in place of oxygen. *Facultative organisms* can perform their functions in either aerobic or anaerobic situations. If the system is aerobic, the facultative microorganism will utilize oxygen, whereas if there is an absence of oxygen (anaerobic system), these microorganisms are capable of using nitrite and other materials in place of the oxygen (see Chapter 4 for nutrient cycles). In some water-treatment methods the unique properties of aerobic and anaerobic microorganisms are utilized to purify wastewater (see Chapter 9).

human implications

The microbial populations in water change and fluctuate in response to the source of the water in which they are found, as well as in response to the composition of the various materials found in the water. The origin of these microorganisms is generally the result of surface runoff, precipitation that carries microorganisms attached to dust particles to earth and/or the addition of wastewater to waterways, bays, lakes, and so on. Generally, the microorganisms that are derived from runoff from uncontaminated terrestrial areas and from precipitation are not harmful to man.

The microorganisms derived from wastewater may, on the other hand, be harmful. These microorganisms are termed *pathogens.* Generally, wastewater contains both pathogenic and nonpathogenic forms. The danger of disease arises when wastewater enters water supplies intended for human use.

Diseases commonly transferred in this manner are those of enteric origin (those caused by microorganisms found in the intestinal tract). Microbial pathogens found in excretory matter include *Shigella,* the agent of bacillary dysentery; *Salmonella,* the bacterial agent of typhoid fever; *Vibrio,* the agent of cholera; *Endamoeba,* a protozoan responsible for amebic dysentery; and two viral agents, polio and hepatitis.

The threat of disease by contaminated water supplies is of prime concern; thus constant testing of water supplies is essential. Since the routine examination of drinking water for the presence of intestinal pathogens is difficult and costly to perform, water supplies are generally tested for the presence of nonpathogenic forms such as *Escherichia coli* and *Streptococcus fecalis.* These organisms are always found in the intestinal tract but are not routinely found in soil and water. Thus when they are detected in water supplies it is assumed that the water is contaminated. These organisms, therefore, are used as indicators, since their presence indicates the possibility that other intestinal microorganisms may also be present—the intestinal pathogens. All qualitative bacteriological testing of water is based on the identification of these indicator species. Table 2-1 summarizes the major microbial indicators. The reader is referred to *Bergey's Manual of Determinative Bacteriology* for specific classification methods.

TABLE 2-1

microbial indicators of water quality

Indicator	*Specific Organisms*	*Use*
Fecal coliforms	*Escherichia coli*	Indicator of fecal contamination by warm-blooded animals

TABLE 2-1 (cont.)

Indicator	Specific Organisms	Use
Total coliforms	*Escherichia aerobacter*	Indicator of coliforms originating in soil and intestines of warm-blooded animals
Fecal streptococci	*Streptococcus faecalis* *S. durans* *S. faecium* *S. bovis* *S. equinus*	Used to judge the probability of human or livestock contamination source; used in conjunction with fecal coliforms
Algae	Polluted water: *Oscillatoria, Spirogyra, Anabaena,* others Clean water: *Ulothrix, Galothrix, Cyclotella, Diatoma, Chrysococcus,* others	Large accumulations of polluted-water forms indicate advancing eutrophication
Sulfur bacteria	*Thiobacillus* Other colorless, green and purple sulfur bacteria, purple sulfur bacteria	Accumulations indicate sulfur-bearing wastes, particularly containing hydrogen sulfide
Iron bacteria	*Gallionella* *Ferrobacillus ferrooxidans* Others	Accumulations in natural waters indicate iron-bearing wastes
Standard plate count	Various	Empirical techniques for determining general bacterial density; identification permits judgment concerning relative importance of parasites, pathogens, and beneficial saprophytes

SUGGESTED READINGS

Brock, T. 1970. *Biology of Microorganisms.* Englewood Cliffs, N.J.: Prentice-Hall, Inc.

Buchanan, R.E., and N.E. Gibbons, editors. 1974. *Bergey's Manual of Determinative Bacteriology,* 8th ed. Baltimore: Williams & Wilkins Company.

Pelczar, Michael J., and Roger D. Reid. 1972. *Microbiology*. New York: McGraw-Hill Book Company.

Stanier, R., M. Doudoroff, and E. Adelberg. 1970. *Microbial World*. Englewood Cliffs, N.J.: Prentice-Hall, Inc.

QUESTIONS

1 / Define symbiosis.

2 / List thc differences among commensilism, mutualism, and parasitism.

3 / Why is saphrotism considered an indirect microbial relationship?

4 / What is the difference between parisitism and predation?

5 / What are the differences between anaerobic and aerobic decomposition?

CHAPTER 3

the hydrologic cycle

The hydrologic cycle is the process whereby water is converted from its liquid or solid state into its vapor state. As a vapor the water is capable of traveling considerable distances from its source prior to recondensing and returning to earth as precipitation. Thus the hydrologic cycle is a complex, interrelated system involving the movement of atmospheric, surface (marine and fresh), and groundwater throughout various regions of the world. It is the hydrologic cycle that is solely responsible for the world's precipitation, and it is this precipitation, falling on the terrestrial and surface freshwater environments, that is the sole source of the earth's supply of fresh water (Fig. 3-1).

The hydrologic cycle may consist of either a long or various short cycles. In the short cycles, water may evaporate from either marine or freshwater systems, condense almost immediately, and return as precipitation to the same system. Another variation of a short cycle is the precipitation and subsequent evaporation of water from land surfaces, followed by its condensation and return as precipitation to the land, followed by reevaporation, and so on.

In the long cycle the major source of water vapor is the world's oceans, which contain 97.3% of the earth's waters. In this cycle a portion of the water evaporates and forms clouds that move inland. The water vapor then cools and returns to earth as precipitation. It is estimated that only 0.007% of the oceanic water is distributed to terrestrial areas annually. This water will ultimately return to the oceans through river and groundwater flow. Since precipitation may occur close to the source of initial evaporation or thousands of miles away, the water may remain in the vapor state for variable times (a few hours to a few weeks). The average residence time for water to remain in the atmosphere is considered to be 10 days.

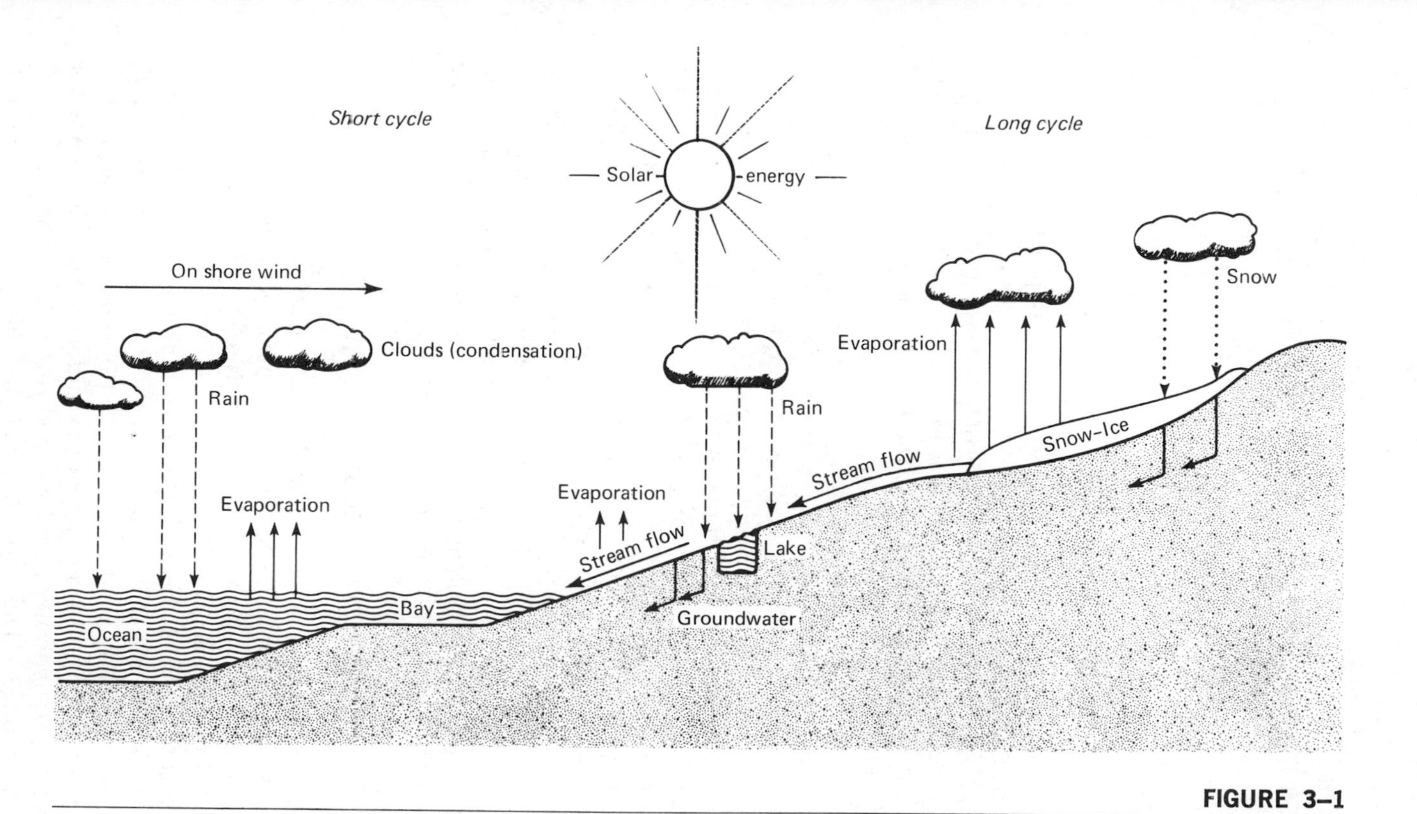

FIGURE 3–1

The hydrologic cycle.

stages in the cycle

evaporation and condensation

Water vapor, evaporated primarily from the oceans, tends to condense around minute particles, termed *nuclei,* which are suspended in the atmosphere. The nuclei generally consist of small particles of organic material (spores, pollen, etc.), fine mineral particles, volcanic ash, and the like. Dust and smoke particles from industrial sources and automotive exhausts may also serve as nuclei and are a major factor in contributing to the contamination of rainwater. This is due to the fact that airborne particles (such as lead) readily dissolve in the newly condensed atmospheric water, and these materials are then returned to earth along with the rainwater.

Initially, the condensed liquid is in the form of extremely small droplets [less than 0.04 millimeter (mm) in diameter]. Because of their small size, their rate of fall is negligible, and they are retained in the atmosphere as clouds. Ultimately, however, this vapor forms precipitation in response to one of three major factors. The largest amount of precipitation forms when masses of warm, moist air move into regions of cold air. This results in a rapid condensation of the vapor and subsequent precipitation. The second factor arises during periods of warm weather, when air, warmed at the earth's surface, decreases in density and rises into overlying cold air, bringing about condensation of the water vapor. The third mechanism involved in condensation occurs with the cooling of air masses as they move over high mountains. In each case the cooling of warm, moist air is responsible for the condensation and precipitation of rainwater.

runoff, stream flow, and infiltration

When rain falls on a terrestrial area, a portion of the water is caught by vegetation. This process is termed *interception* and this water is readily reevaporated, since there is a large surface area that is exposed to wind action. The remainder of the water falls to earth, and a portion sinks into the soil surface by a process known as *infiltration.* The portion not infiltrated into the soil, termed *surface runoff,* flows over the surface and is discharged (in undeveloped areas) into streams. The water entering a stream by surface runoff plus the water entering via groundwater flow is termed *runoff*. Thus the terms "runoff" and "surface runoff" are different and distinct. Surface runoff equals precipitation minus the water lost by interception and infiltration. Runoff, on the other hand, is generally synonymous with *stream flow* and is the sum of the surface water plus the groundwater that enters a stream.

Infiltration is also distinct from groundwater per se because when the water percolates into the earth's surface, different portions will follow three distinct pathways. A portion of the water will function as interflow, because the presence of impermeable lower sediments will prevent deep penetration of this water. Consequently, this segment will flow just below the soil surface and discharge into streams. Another portion will remain above the water table in an area termed the *zone of unsaturated flow*. Both of these portions of water are considered to be in the *zone of aeration*. The third portion will percolate down into the groundwater table (Chapter 5) and will, eventually, be discharged into streams.

evapotranspiration

A large amount of the precipitated water is converted to vapor by evaporation and/or transpiration. *Evaporation* is the process whereby molecules of liquid water (at the surface of a water body or in moist soil) gain sufficient energy to leave the liquid state and enter the vapor state (Chapter 1). The energy absorbed by the water in this process is stored within the vapor—hence the rather large amount of heat contained in the moist air.

Transpiration is the process whereby terrestrial and emergent aquatic vegetation release water vapor to the atmosphere. In the process of photosynthesis (Chapter 4) all plants take in liquid water and carbon dioxide and, by a complex series of reactions, convert these materials to carbohydrate, oxygen (gas), and water vapor. Submergent aquatic vegetation (plants growing completely beneath the water surface) also releases water and oxygen as vapors but, in these cases, the oxygen and water vapor produced and released into the surrounding liquid water will immediately form hydrogen bonds with the liquid water. The water vapor will be converted to its liquid form, while the oxygen remains in the water column as dissolved oxygen. Since the leaves of both terrestrial and aquatic emergent vegetation are in air, the water vapor and oxygen, when released by these plants in photosynthesis, will remain in the vapor state because there are insufficient water molecules in the immediate vicinity to form hydrogen bonds and place the water vapor and oxygen in solution.

The water converted from liquid to vapor in transpiration is considerable, and the ability of plants to remove water from both soil and aquatic systems cannot be underestimated. Transpiration by emergent vegetation is a major factor in the "drying up" of lakes (Chapter 4).

Groundwater, unless it is within a few feet of the surface, is not evaporated. Rather, the portion that is converted to vapor is transpired by plants. In most regions the water lost by evaporation cannot be measured separately from the water lost by transpiration. Consequently, the two are considered together as *evapotranspiration.*

groundwater

The most lengthy portion of the hydrologic cycle is completed when groundwater is returned to the earth's surface. The return may occur by springs, transpiration, or by artificial means (Chapters 5 and 8).

Any natural surface discharge of sufficient water that will flow as a small rivulet is termed a *spring,* while a smaller discharge is called *surface seepage* (Fig. 3-2). Groundwater may also be discharged as subaqueous springs below the surface of lakes, rivers, and marine systems. Springs are generally classified on the basis of their magnitude of discharge, as summarized in Table 3-1. Three major factors

TABLE 3-1

classification of spring discharge

Magnitude	*Discharge*
First	Greater than 2.83 m^3/s
Second	0.283–2.83 m^3/s
Third	2.83–283 liters/s
Fourth	2.83–6.31 liters/s
Fifth	6.31–0.631 liters/s
Sixth	63.1–631 ml/s
Seventh	7.9–63.1 ml/s
Eighth	Less than 7.9 ml/s

affect the magnitude of a spring: the land area accepting the rainfall and thus contributing water to the subsurface system, the permeability of the soil and subsoil, and the quantity of water entering the system (the amount of *recharge*). The use of water, in developed areas, prior to recharge affects the quality and quantity of the water entering the subsurface systems and is discussed in Chapter 8.

FIGURE 3–2

Surface seepage. (*Photo: G. Marquardt*)

All the world's waters are, therefore, interconnected by the hydrologic cycle. Oceanic water is evaporated and a portion is carried as water vapor over terrestrial regions, where it returns to earth as precipitation. This water may then be evaporated or transpired, or it may enter streams, rivers, lakes, or the groundwater system. Regardless of the pathways that the water may take, it is the sole source of fresh water and, therefore, the sole source of domestic and industrial water. Ultimately, and generally only after use and reuse by man, this water will return to the sea—only to be evaporated and to reenter the hydrologic cycle.

SUGGESTED READINGS

Davis, Stanley N., and Roger J. M. DeWiest. 1966. *Hydrogeology*. New York: John Wiley & Sons, Inc.

Kuenen, P. H. 1963. *Realms of Water*. New York: John Wiley & Sons, Inc.

Leopold, Luna B. 1974. *Water: A Primer*. San Francisco: W. H. Freeman and Company.

QUESTIONS

1 / Define the hydrologic cycle.

2 / Distinguish between long and short cycles.

3 / Explain how air pollution may lead to water pollution.

4 / Explain evapotranspiration.

5 / Distinguish between runoff and surface runoff.

CHAPTER 4

freshwater systems

Although freshwater environments occupy only a small portion of the earth's surface in comparison with marine systems, they are extremely important in water management since they are the cheapest and most convenient source of consumptive and industrial water. These systems simultaneously provide an easy, economical method of the disposal of waste.

general characteristics

Although freshwater systems are commonly divided into two major categories—the *lentic,* or *standing-water systems* and the *lotic,* or *running-water systems*—both systems share certain chemical, physical, and biological characteristics. The major phenomenon that serves to relate all systems is the biological food chain, which encompasses physical, chemical, and biological principles. Since the food chain is basic to all systems, terrestrial, marine, or freshwater, it will be discussed in detail here and merely referred to in later chapters.

The *food chain* refers to the transfer of energy (in the form of food) from its source in green plants through a series of "higher" organisms, the animals inhabiting a given system. The energy is transferred through the food chain by the process of grazing or predation. Since the plants are the original source of the food energy, they are termed the *producers,* or *autotrophs.* Animals feeding directly upon the plants are termed *herbivores,* or *primary consumers,* whereas *predators* are termed *secondary, tertiary,* etc., *consumers,* depending upon the actual source of their food. A well-managed fish pond will serve to illustrate a typical food chain. In this system the plants, generally phytoplankton (single-celled plants), are fed upon by the primary consumers, the zooplankton. The zooplankton, in turn, serve as a food source for bloodworms, or chironomids, which occupy the position of secondary consumers. The bloodworms are the food source of small fish such as bluegills (tertiary consumers) and are, in turn, preyed upon by bass (quaternary consumers). Thus the food manufactured by the plants passes from animal to animal and all food chains are, therefore, dependent on the productivity of the green plants to supply the necessary food to maintain the system.

In order to perform the vital function of photosynthesis, plants require two basic materials: a source of inorganic materials (phosphate, nitrate, and CO_2) for plant growth and metabolism; and sunlight, to provide the necessary energy to convert these materials into a form suitable for use by the consumers. During this process oxygen is also produced and released into the system to be used by the animals in respiration. The animal component of any system cannot use these inorganic materials directly. Rather, they must be taken in by the plants and converted into the proper form prior to use by the consumers. Thus it is only through the mediation of the producers (green plants) that these materials are converted into their proper form and then stored in this form, as plant tissue, prior to consumption by the animals.

Since sunlight is such an important factor in the productivity of these systems, it is customary to divide both marine and freshwater systems into definite zones on the basis of the depth of sunlight penetration. This is primarily a biological method of zonation. The various wavelengths of light are rapidly filtered out as they pass through a water column. Based upon the degree of light penetration, it is possible to divide a system into certain zones (Fig. 4-1). The upper zone of sunlight penetration is termed the *euphotic zone*. Beneath this is a

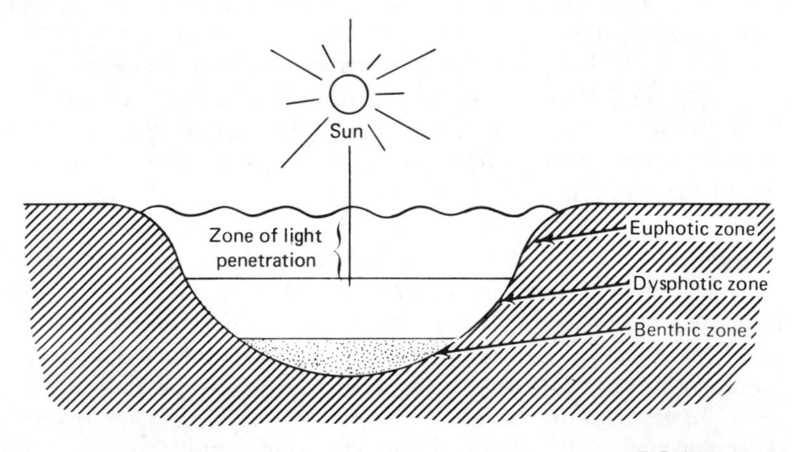

FIGURE 4–1

Lake zonation on the basis of sunlight penetration.

zone of perpetual darkness, the *dysphotic zone,* which extends from the bottom of the euphotic zone to the bottom. The bottom sediment underlying the dysphotic zone is termed the *benthic zone* or the *regeneration zone.*

According to this type of classification, shallow lakes, streams, and rivers would not have a dysphotic zone, since they would be shallow enough to allow for sunlight penetration down to the benthic zone. In addition, the position of both the euphotic and dysphotic zones in deep systems generally shifts with environmental conditions. For example, during periods of heavy rainfall, when sediment is washed into the system, the physical presence of this suspended sediment would tend to block the sunlight, causing a reduction of the euphotic zone and an increase of the dysphotic zone.

The productivity of all systems is governed, primarily, by both the available nutrients present and the amount and degree of sunlight impinging on a given system. Although light relations and nutrients are considered separately in this text, it is to be stressed that all systems require both light and a sufficient supply of nutrients. In the absence of either, the plants will be eliminated and productivity will cease.

Sufficient sunlight is vital to the productivity of all systems, since green plants require sunlight as the energy source to trigger the production of food and oxygen by photosynthesis. In photosynthesis the producers (plants) take in carbon dioxide, liquid water, and the necessary nutrients and, by a complex series of biochemical reactions, convert these materials to carbohydrate (CHO_n), gaseous oxygen, returning the water to the system as water vapor. The process is summarized by the reaction (unbalanced equation)

$$CO_2 + H_2O \rightarrow CHO_n + O_2 + H_2O$$

In aquatic systems the water vapor and gaseous oxygen go immediately into solution since the water molecules form intermolecular attractions (hydrogen bonds) with the gaseous oxygen as well as with the water molecules evolved in photosynthesis.

The reverse reaction ($CHO_n + O_2 \rightarrow CO_2 + H_2O$) is termed *respiration* and is carried on by all the animals as well as the plants inhabiting a given system. Although the plants both photosynthesize and respire (using a portion of the oxygen and carbohydrate produced in photosynthesis), during the day they produce more oxygen and carbohydrate than they use. The end result, therefore, is a net gain of both oxygen and carbohydrate. The animal component carries on no photosynthesis but merely respires both day and night. They obtain their energy by feeding on the plants (or, if higher consumers, on primary, secondary, and so on, consumers), while the oxygen that they use in respiration was produced by the plants in photosynthesis. At night, however, both plants and animals respire, and since there is no sunlight to trigger photosynthesis, no oxygen is produced by the plants. Consequently, oxygen levels decline while CO_2 levels increase in the water column. If oxygen concentrations in a typical system were measured hourly for a 24-hour period and the results graphed (see Appendix III for methods), the curve shown in Fig. 4-2 would be obtained. This curve indicates that oxygen production, and thus photosynthesis, increases and reaches a peak shortly before noon. As the sunlight decreases, so does photosynthesis and the production of oxygen. A typical curve for CO_2 levels during the same period would show the reverse concentrations (Fig. 4-2).

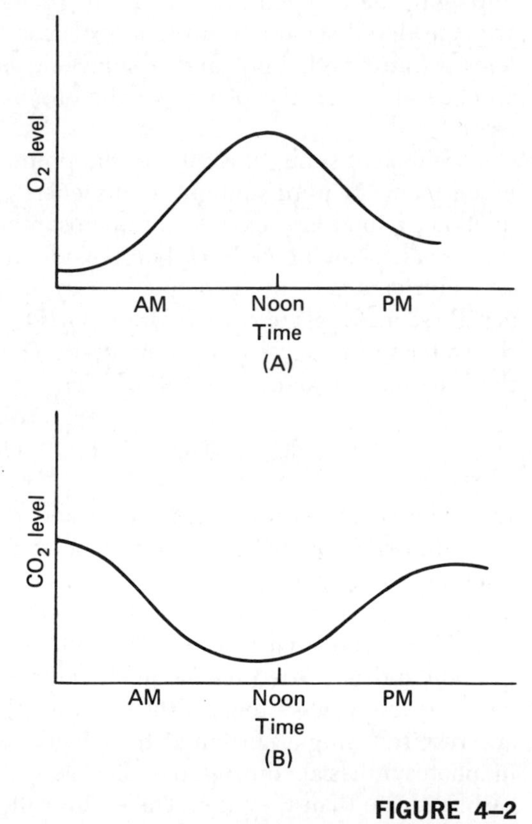

FIGURE 4–2

The relationship of sunlight to O_2 and CO_2 levels.

The intensity of light will, therefore, control the entire system by influencing the primary productivity of the plants. It is found, however, that at high light intensities photosynthetic activity is actually decreased. If the relative photosynthesis as well as the intensity of light impinging upon the system is measured (Fig. 4-3) in terrestrial, marine, and freshwater systems, it is found that there will be a linear increase in photosynthetic rate up to an optimum or saturation intensity, followed by a rapid decrease at higher light intensity. The saturation intensity varies from only a few hundred foot-candles for some species to several thousand foot-candles for most species. This is due to the fact that at high intensities photooxidation of photosynthetic enzymes will apparently reduce the photosynthetic rate of the plants. It is common, therefore, when measuring oxygen levels or photosynthetic rates of aquatic communities to obtain definite midday depressions during periods of full sunlight.

In typical aquatic and marine systems the euphotic zone is considered to be both the zone of sunlight penetration and the zone of oxygen and food production. It is in this zone that the plants and the primary consumers would be found, at least during the periods of active photosynthesis and grazing. The other consumers (secondary, tertiary, etc.) would be found in all three zones, depending upon the availability of their food source.

As noted previously all systems require two major chemical components for the growth, reproduction, and general metabolic processes of the organisms that inhabit those systems. These components, termed *essential elements,* are phosphorus and nitrogen. Plants require phosphorus and nitrogen in their dissolved form as inorganic phosphate (termed *orthophosphate* and abbreviated $o\text{-}PO_4$) and inorganic nitrate

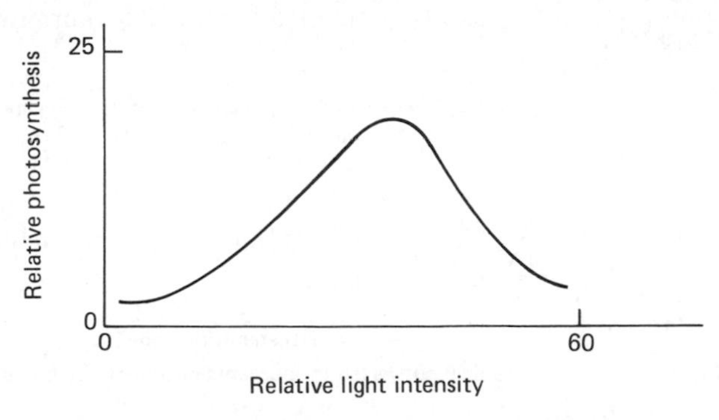

FIGURE 4–3

The relationship of light intensity to photosynthesis.

(NO_3). Although animals also require phosphorus and nitrogen, they require it in its organic form as organic phosphorus and organic nitrogen (generally obtained in the form of amino acids or protein by grazing on plants or predation on other consumers). Should animals attempt to ingest phosphate or nitrate in its inorganic form, it would be totally unusable and immediately excreted.

Phosphorus and nitrogen tend to cycle (Figs. 4-4 and 4-5) in systems, passing from plant to various animal communities, then into their inorganic forms, and then back into the plant components of the system. The discussion of these cycles is conveniently begun with those plants which will take in the inorganic materials. During the process of photosynthesis, the plants convert o-PO_4 and NO_3 into their organic forms such as nucleic acids (DNA, RNA), amino acids, and proteins, which contain various ratios of phosphorus and nitrogen in their molecular structures. These materials are utilized by the plants in their metabolic process and stored as new plant tissue during the growing season. Herbivores (primary consumers) obtain their phosphorus and nitrogen by feeding directly on the plants in the system, while the higher consumers obtain their phosphorus and nitrogen by predation. At each step in the food chain some of the material is lost to the next link by death and/or excretion. In either case the phosphorus and nitrogen contained in the excretory matter or the bodies of the plants and/or animals not consumed will sink through the water column and enter the benthic zone. Although bacterial decomposition presumably begins from the moment of excretion or at death, the majority of the decomposition occurs in the benthic sediments (regeneration zone). It is in the regeneration zone that bacterial action (see Chapter 2 for details) converts the organic nitrogen and phosphorus into their inorganic form. It is to be noted (Fig. 4-4) that phosphorus is readily converted into its inorganic form. Organic

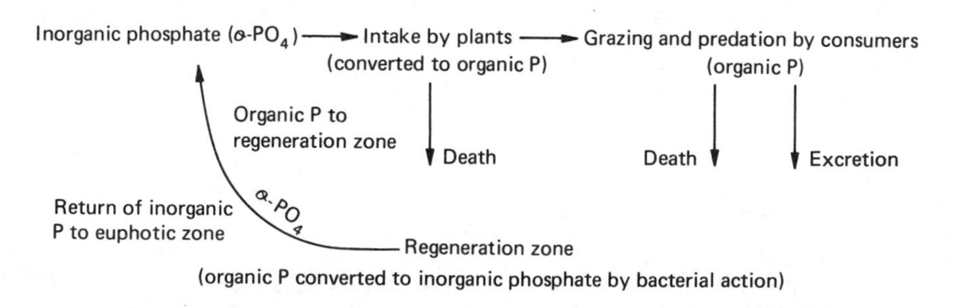

FIGURE 4–4

The phosphorous cycle.

nitrogen (Fig. 4-5), however, converts less rapidly and is broken down in a three-step bacterial process by a series of decomposer organisms, each specialized to accomplish a specific step in the decomposition process. Initially, the organic nitrogen in the form of amino acids and protein is converted to ammonia to nitrite (NO_2) and finally, through the mediation of yet another group of bacteria, the nitrite is converted to nitrate (NO_3), the form required by most species of plants. The decomposition from protein and amino acids to nitrate provides an energy source (food) for the bacterial species that perform this breakdown; the reverse process (NO_3 to protein) requires energy from sunlight, and this process is accomplished during photosynthesis.

Plants inhabiting any system require nitrogen and phosphorus in certain ratios. The ratios required by plants in both terrestrial and freshwater systems are extremely variable and thus are dependent upon the specific populations of plants that inhabit a particular system. Since, however, plant communities do require not only phosphorus and nitrogen but require these materials in certain ratios, it is the availability of these nutrients in their proper ratio that becomes

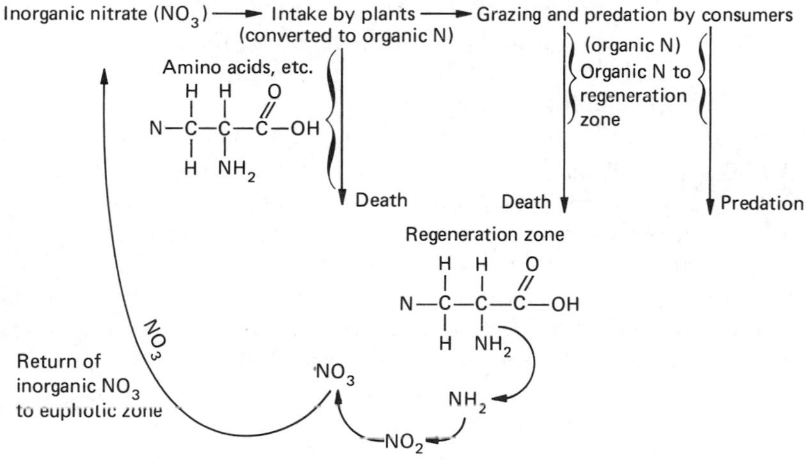

Inorganic nitrate (NO_3) → Intake by plants (converted to organic N) → Grazing and predation by consumers

Amino acids, etc.

(organic N) Organic N to regeneration zone

Death

Death

Predation

Regeneration zone

NH_2

NO_2

NO_3

Return of inorganic NO_3 to euphotic zone

FIGURE 4–5

The nitrogen cycle, depicting the conversion (regeneration) of an amino acid to the inorganic form (NO_3). This conversion occurs in three distinct steps—each mediated by a specific group of microorganisms. In step 1 the NH_2 is removed from the parent molecule. In step 2 the Hs are removed and two Os added, forming the nitrite. In step 3 the nitrite (NO_3) is formed by the addition of another O.

important from the viewpoint of productivity. Since these ratios are so variable in freshwater systems, it is preferable to cite a hypothetical example. Assume that in a given freshwater system it has been determined that the proper phosphate/nitrate ratio is 1:8. This would indicate that (at least in this system) the plants require one phosphate for every eight nitrates, and that during the periods of active plant growth and metabolism, the plants would be removing phosphate and nitrate from the water in a 1:8 ratio and tying it up, as plant tissue, in this ratio. The consumers, too, would be obtaining the same phosphorus/nitrogen ratio. When these organisms died and fell to the regeneration zone, the decomposers would, ultimately, return this material to the water column in this ratio.

In a system in which the producers require a 1:8 ratio of phosphate to nitrate, should the ratios become disturbed (1:16, for example) the plants would tend to remove 8 nitrates and all (1) phosphate. The remaining 8 nitrates could not be utilized, because there would not be sufficient phosphate in the system. In this case the factor limiting further plant growth would be phosphate. If the situation were reversed and there were a 2:8 ratio, the limiting factor would be the lack of additional nitrates. In this case the plants would remove all the nitrate, but, since the nitrates were present in the lesser ratio, they would only be able to utilize one of the phosphates.

In many studies it is advisable to not only measure the inorganic phosphate and nitrate present in a given system (see Appendix III for methods), but also to determine the amount of nitrate, ammonia, and organic phosphate. By performing these additional analyses, not only can the productivity at the time of sampling be determined, but also the potential productivity in a given system can be inferred

In the majority of freshwater systems it is found that phosphate is generally the limiting factor. This is due to the presence of blue-green algae, which are common in most freshwater systems. These forms have the ability to fix atmospheric nitrogen, thus increasing the concentration of nitrate in the water column. Apparently the normal phosphate input into these systems from the weathering of phosphatic terrestrial materials is insufficient to counterbalance the amount of nitrogen input by the blue-green algae.

As noted previously, both lentic and lotic systems have the same basic mechanisms of nutrient cycling and sunlight relationships as noted above. Beyond these commonly shared characteristics, however, the systems vary greatly and should be considered as separate and discrete entities.

the lotic environment

The *lotic environment* consists of all inland waters in which the entire water body continually moves in a definite direction. Thus rivers, streams, and brooks are considered to be lotic environments (Fig. 4-6). These systems generally originate from precipitation that falls on the earth's sloping surface and flows downhill as *sheet wash.* Sheet wash

FIGURE 4–6

The lotic environment. (*Photo: G. Marquardt*)

tends to accumulate in low areas and to form intermittent rivulets, which eventually flow into streams. As the water flow continues, it will tend to erode the bed of the stream, causing it to intercept the groundwater table (Chapter 5) and become a permanent stream. As additional streams are formed and eventually meet, the resulting length, depth, and size of the system will warrant classification as a river. The rivulet–stream sequence can still be observed upstream, but in an ever-changing position, since erosion will cause an upstream migration of these features. The migration will continue until halted by a natural feature such as a drainage divide. A direct result of this erosion-induced geological migration is a continual migration of biological habitats. In other words, erosion will cause the headwaters to "migrate" upstream, followed by a similar "migration" of the environmental conditions characteristic of rivulets, streams, and rivers. The animal and plant communities that occupy each of these habitats must also move upstream, to remain in the same habitat (which is in a constantly changing position as a result of erosion), or must adapt to the gradually changing conditions brought about by the erosion caused by the running waters. In natural, undisturbed systems these changes are very slow and allow ample time for the organism to adapt to the changing conditions.

Since lotic systems tend to be shallower and narrower than lakes, there is generally a greater proportion of water exposed to land surfaces. Consequently, streams are more intimately associated with and affected by the terrestrial environment than are most lentic systems. In addition, lotic systems are more dependent on the surrounding land areas for a large proportion of their nutrient supply. The rapidly flowing waters of these systems tend to carry phytoplankton rapidly downstream and to discourage the growth of rooted plants in the portions exposed to rapid stream flow. In typical stream habitats both nutrient levels and plant materials tend to fluctuate seasonally in response to the materials entering from the terrestrial environment (in the form of windblown leaves and the like) as well as in response to nutrient fluctuations of their source waters. For example, if a large portion of a river's water originates from a groundwater source that underlies an extensive agricultural area, it would be expected that the groundwater (and thus eventually the river) would be high in nutrients during the spring and summer months when the growing season (and fertilization) is at its peak. During the winter months, with no fertilization, the nutrient levels of the groundwater would decrease, leading to a reduced nutrient level in the river. In lotic systems, since they are running waters, the nutrients tend to cycle from habitat to habitat vertically as they are carried downstream. This is in direct contrast to lentic systems, in which the nutrients tend to cycle "in place" (Fig. 4-7) and move vertically from the regeneration zone up to the euphotic zone and then downward again as organically bound debris (see p. 40).

In both terrestrial and lotic systems, therefore, the nutrients are said to exhibit a "downhill" tendency. This is the tendency for organically bound phosphorus and nitrogen (in the form of leaves, plant and animal fragments, etc.) from the terrestrial environment to enter streams by means of a variety of processes. Once this material is in the stream, it may be used directly by herbivores entering into the consumer portion of the food chain, or, through bacterial action, it may begin to decompose. During decomposition, however, it will be carried downstream by the current and become available in a different location. These materials will be taken in by plants inhabiting this downstream position, converted into their organic forms, and become available to the animals of that area. Eventually, through death or excretion, these materials will enter the regeneration zone (in all likelihood far downstream) to be returned to their inorganic form. Consequently, the nutrients will cycle on the way downstream and will become available to many different communities of plant, animal, and bacteria before reaching the marine environment, where they will enter into the marine food chain (Chapter 4).

Since the majority of lotic systems are both shallow and turbulent, they tend to warm and cool uniformly in response to atmospheric temperatures. The organisms that inhabit these areas have a rather wide temperature tolerance. The shallow nature of these systems permits large surface areas to be exposed to air, and the oxygen levels of lotic systems are generally high, since the water is in constant motion. Organisms that inhabit this environment have a narrow tolerance to oxygen fluctuations and are very sensitive to the input of any materials that would decrease the oxygen concentrations of the surrounding waters. Since these systems are shallow, sunlight is generally able to penetrate to the bottom. The entire water column is, therefore, in the euphotic zone and phytoplankton may be found at all depths. Rooted vegetation may also become established on the bottom.

the lentic environment

The *lentic environment* consists of all inland waters in which the water is not continually flowing in a definite direction. The water in these systems is essentially standing, although some water motion may occur as a result of wind-driven waves and/or in the vicinity of inlets and

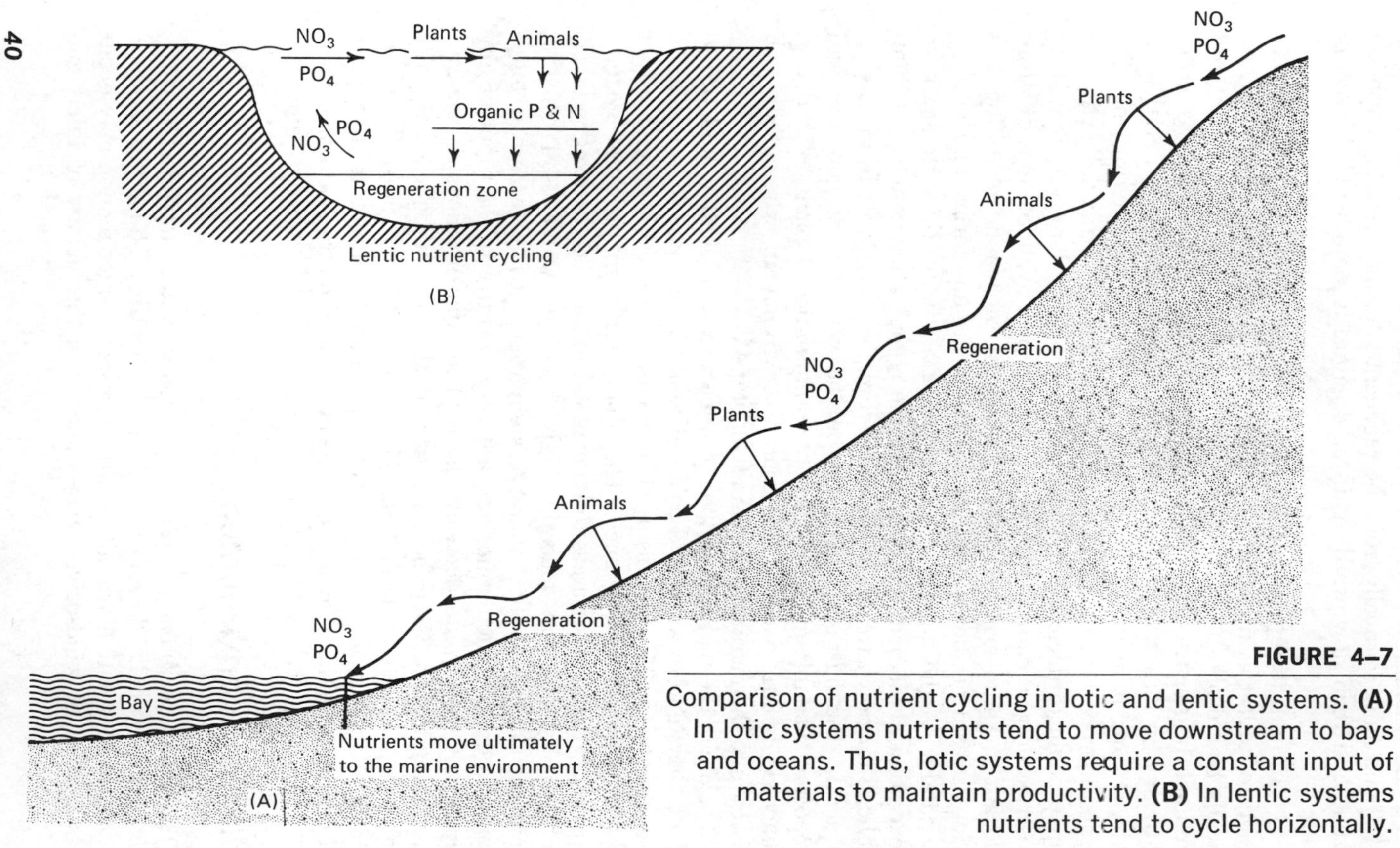

FIGURE 4–7

Comparison of nutrient cycling in lotic and lentic systems. **(A)** In lotic systems nutrients tend to move downstream to bays and oceans. Thus, lotic systems require a constant input of materials to maintain productivity. **(B)** In lentic systems nutrients tend to cycle horizontally.

outlets. From a long-term geologic viewpoint, however, these inlets and outlets are considered to be temporary impoundments that will in time disappear, converting the system to a flowing-water environment. Typical lentic environments are considered to be lakes, ponds, and marshes.

In lentic systems the aging process is just the reverse of that described for lotic systems. Whereas streams tend to get wider and deeper as they age, lakes tend to get shallower and the banks extend into what was originally open water. Natural filling is generally due to wind-blown materials (sand, leaves, etc.) entering the system, sediment input by streams, and terrestrial runoff and aquatic plant and animal debris. Not all lentic environments become shallower by this type of filling alone, however. In many cases outlets may widen and deepen, causing an increased outflow of water with a subsequent lowering of the water level.

Lakes are generally classified, on the basis of their depth and nutrient levels, as oligotrophic, eutrophic, or senescent. An *oligotrophic lake* is considered to be in its young stage. These lakes are very deep, with a sand or rock bottom, and are low in nutrients. Because of the paucity of nutrients, both plant and animal life are low. *Eutrophic lakes* are middle-aged systems. They are relatively shallow in comparison to oligotrophic systems, with a silty or mud bottom, and have sufficient nutrients to support a large population of plants and animals. A *senescent lake* is in the oldest stage of development. The bottom sediments in senescent systems consist of a thick layer of organic silts and/or muds, nutrient levels are high, and the system is very shallow. There is a large percentage of rooted emergent vegetation growing throughout the system. Terrestrial or marsh vegetation tends to grow along the banks and into the lake itself over the root mat.

In addition to the general method of zonation discussed previously, oligotrophic and deep eutrophic lakes can be divided into zones on the basis of the temperature and density of the water at various depths. This is primarily a physical method of classification, since it deals with the density relationships of the water column (see Chapter 1).

On the basis of temperature (and thus, indirectly, density) lakes can be divided into three distinct zones: the epilimnion, the thermocline, and the hypolimnion (Fig. 4-8). The *epilimnion* is the upper zone of gradual temperature change. Below the epilimnion is a zone of rapid temperature change, termed the *thermocline*. In order for this region to meet the established criterion for classification as a thermocline, the temperature must change by at least 1°C for every meter of depth. If the temperature changes at a lesser rate relative to

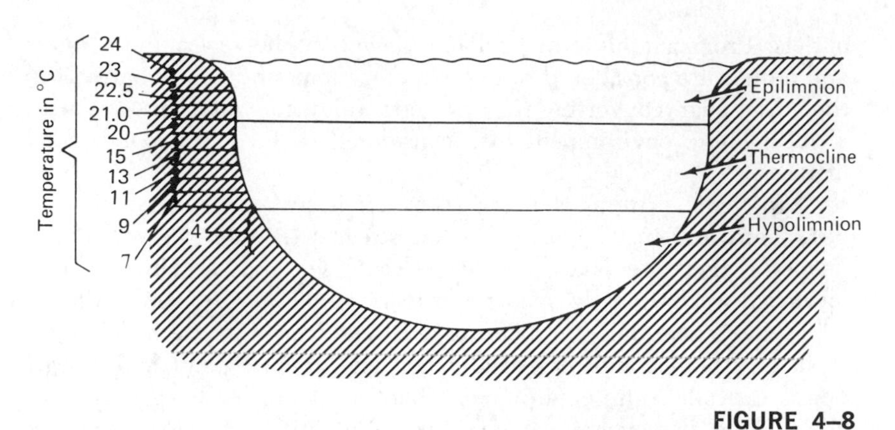

FIGURE 4–8

Typical lake stratification—summer.

depth, these waters would be classified as a portion of the epilimnion. Below the thermocline there is an area of water that is constantly at a temperature of 4°C, termed the *hypolimnion.* Since the waters of the hypolimnion are at 4°C they are at their maximum density and will be found at the bottom layer in any system. Comparing the relative water temperatures and density variations seasonally, it is noted that in summer the warmer water will be located in the upper epilimnion and the colder, denser, water will be located in the hypolimnion. In the winter the colder, less dense water at temperatures below 4°C will be found in the epilimnion, and the denser water at 4°C will be located in the hypolimnion. In either case, the less dense water is in the epilimnion and the dense (4°C) water is always found in the bottom waters of the hypolimnion. Figure 4-9 summarizes the seasonal stratification.

Comparing the biological method of classification (p. 30) with the physical method of classification, it is to be noted that, although a portion of the epilimnion (physical classification) will always be in the euphotic zone (biological classification), these areas do not always coincide. In reality, it is the exception rather than the rule when these two areas do actually coincide. For example, in a very clear lake the euphotic zone may extend down into the thermocline, whereas in a lake subject to a high degree of surface runoff, only a small portion of the epilimnion may fall into the euphotic zone. The major possibilities are summarized in Fig. 4-10.

As a result of the polarity of water, oligotrophic (and deeper eutrophic) lakes undergo definite physical changes, and the chemical and biological characteristics of the lake also undergo definite, predictable seasonal changes. If temperatures at various depths in typical, tem-

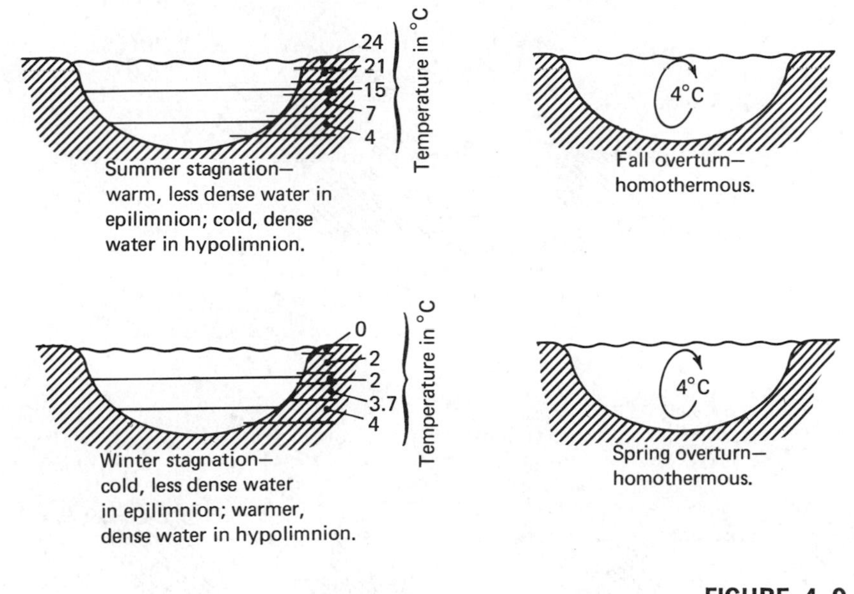

FIGURE 4–9

Schematic illustration of the seasonal occurrences in an oligotrophic lake.

perate-zone oligotrophic lakes are measured (see Chapter 11 for methods of simultaneously measuring temperature and depth) in midsummer, it is found that the surface waters will be at or close to the atmospheric temperature (e.g., surface to 1 m, $T = 26°C$). As the depth increases the temperature will decrease gradually (e.g., from 1 to 2 m, $T = 25.5°C$; 2 to 3 m, $T = 25.2°C$; 3 to 4 m, $T = 24.5°C$; etc.). This zone of gradual temperature decrease would then be classified as epilimnion. Eventually, the waters of the thermocline would be reached. In this area a rapid temperature decrease with depth would be encountered (e.g., 15 to 16 m, $T = 12°C$; 16 to 17 m, $T = 10°C$; 17 to 18 m, $T = 9°C$; etc.). Beneath the thermocline, in the hypolimnion, the water would be at a constant, uniform temperature of 4°C.

Because of the vast temperature differences, and thus the density variation, the less-dense waters of the epilimnion would tend to be effectively cut off from the hypolimnion, which remains at a temperature of 4°C (the point of maximum density). In other words, the lake is physically stratified into definite, discrete zones on the basis of the temperature–density barrier, which prevents any mixing of the epilimnion and the hypolimnion. The lake at this point is said to be in its summer stagnation.

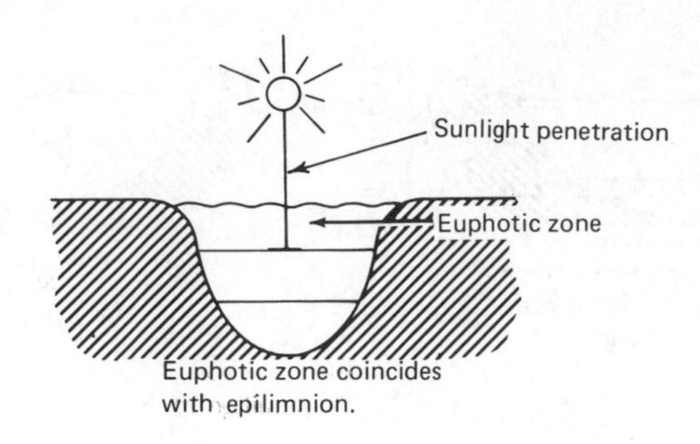

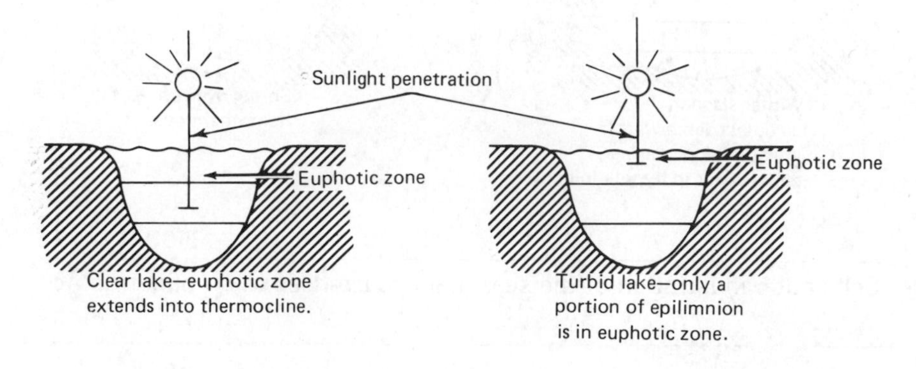

FIGURE 4–10

Lake zonation.

During stagnation periods two simultaneous events occur: nutrient depletion in the epilimnion and oxygen depletion in the hypolimnion. Throughout the summer death of the plants and death and excretion of the animals will continue to add organic material to the regeneration zone that underlies the hypolimnion. The bacteria inhabiting the bottom sediment will continue to break this organic material down and convert it into inorganic phosphate and nitrate, which is the proper form for utilization by plants confined to the euphotic zone (see Figs. 4-4 and 4-5 for cycles). The inorganic phosphate and nitrate will go into solution readily in the waters of the hypolimnion. However, since the lake is stratified, the temperature–density barrier formed by the thermocline prevents these nutrients from reaching the euphotic zone. Thus the nitrate and phosphate are in the proper form for use by the plants, but it is in the wrong place—the hypolimnion. When this occurs the nutrients are said to be spatially unavailable. During a prolonged stagnation these nutrients will build up to large concentrations in the hypolimnion, resulting in a nutrient depletion in the epilimnion. At these times large numbers of plants, with no source of nutrients to carry on their metabolic processes, will die. If this occurs, the animal component of the system, dependent on the plants for the conversion of nutrients into the organic form as well as for the production of oxygen, will also be eliminated.

Oxygen, the production of which is confined to the euphotic portion of the epilimnion, is prevented by the thermocline from reaching the hypolimnion during stagnation periods. The animals that inhabit the hypolimnion as well as the bacteria in the regeneration zone require oxygen for their metabolic processes. The bacteria (aerobic and faculative forms) require oxygen to perform the vital task of aerobic decomposition. As the stagnation period continues, the demand for oxygen in the hypolimnion remains undiminished, while replacement of the oxygen is blocked by the temperature–density barrier of the thermocline. Eventually, during prolonged stagnation periods, the entire hypolimnion may become completely devoid of oxygen (anaerobic), resulting in the elimination of organisms unable to leave the hypolimnion and exist in the epilimnion (coldwater fish, annalids, etc.). Bacterial decomposition is able to continue, but the bacteria carry out this function anaerobically during these periods. The nutrients produced are, as mentioned previously, confined to the hypolimnion and are, therefore, spatially unavailable to the plants.

As the summer comes to a close, a combination of cooler atmospheric temperatures and increased wind action causes the surface waters of the epilimnion to mix slightly. Since cooling occurs at the surface, these waters become colder and denser. Eventually the surface waters become colder and denser than the deeper waters. When this occurs, since they are denser, they will sink to a depth of equal density. This exposes subsurface waters to atmospheric conditions, and these waters will begin to cool, increase in density, and then sink to a depth of equal density. Through this process the entire epilimnion eventually becomes uniform in temperature (homothermous). As the fall temperatures become colder, the epilimnion, by the same process of water cooling, increasing in density and sinking, mixes with the thermocline. In mid- to late fall the entire lake becomes homothermous at 4°C, and there is a complete mixing of water from top to bottom. This is termed the *fall overturn*. At this time inorganic nutrients that have accumulated in the hypolimnion throughout the summer stagnation are returned to the euphotic portion of the epilimnion and become available to the plants. Consequently, the fall overturn is marked by large plankton blooms. Since there is abundant plant life in the lake at this point, there is a corresponding increase in the animal life.

Also during the fall overturn, as a result of the mixing of the entire water column, oxygen-rich water from the epilimnion is circulated down to the thermocline and the hypolimnion. This allows the more efficient aerobic decomposition to take place. During the overturn periods life in the lake is renewed, and the lake is said to "turn over and breathe."

The fall overturn is followed by the winter stagnation period. As atmospheric temperatures continue to cool, a point is reached where the surface waters of the lake (the temperature of the lake prior to the onset of winter stagnation is homothermous at 4°C) begin to cool below 4°C. As cooling occurs, the waters become less dense (Chapter 1 for details). Consequently, the cooler, less-dense surface water tends to remain in place, overlying the denser subsurface water still at 4°C. Eventually a point is reached where the surface waters are cooled sufficiently so that they form a definite epilimnion, which is prevented from effectively mixing with the bottom hypolimnion by the temperature–density barrier of the thermocline. Thus in winter stagnation the cooler, less-dense water is located at the surface and the dense (4°C) warmer water is located in the hypolimnion. This is the reverse of the temperature distribution during summer stagnation; in either case, however, the dense 4°C water is located in the hypolimnion and the less dense (warmer in summer, cooler in winter) water is located in the epilimnion. Although wind action is considerable in winter, the ice cover of the surface-water layer prevents wind mixing and subsequent disruption of the water stratification. It is to be noted, however, that with the smaller temperature variation between the epilimnion and the hypolimnion in winter, the thermocline is not as well established or of as great a volume as during the summer stagnation. The same chemical and biological changes occur during the winter stagnation as have been noted for the summer stagnation.

Predictably, the winter stagnation is followed by the spring overturn. This is brought about by atmospheric temperatures which melt the ice cover and expose the surface waters to wind action and solar heating. As the surface water warms up (above 0°C), the density of the water increases, causing the newly warmed surface water to sink to a depth of equal density. This will expose subsurface water to the atmosphere, and this newly exposed water will be warmed, increase in density, and sink. Eventually, by this repeated process of warming and sinking, combined with wind mixing, the entire lake will become homothermous at 4°C, mix from top to bottom, and enter the spring overturn. During this period, as in the fall overturn, nutrients are brought to the surface and oxygen to the hypolimnion.

As the summer period is entered, the increased atmospheric temperatures warm the surface waters above 4°C. This warming results in the water decreasing in density and again forming layers above the denser, subsurface water. Eventually, the surface water will warm to such an extent (see p. 43 for the typical summer temperatures) that the surface epilimnion will be effectively prevented from mixing with the hypolimnion by the temperature-density barrier of the thermocline. The lake, at this time, will again be in its summer stagnation period (the entire process was summarized in Fig. 4-9).

A lake can be classified as oligotrophic, eutrophic, or senescent, depending on its depth, nutrient levels, and biological composition. The normal sequence is for an oligotrophic lake to age gradually to its eutrophic stage and then to its senescent stage. This aging occurs by two simultaneous processes: filling with terrestrial and aquatic debris; and cutting down of the outlet stream bed, causing a greater volume of water to flow out. Both processes lead to a lowering of the lake surface. Determining the age of a lake is rather inexact, as there is a large variation in the physical and biological characteristics of lentic systems. In general, a eutrophic stage is reached when the nutrients are at a suitable level to support a large population of plant and animal communities and the hypolimnion is consistently smaller than the epilimnion. As a lake ages, the plankton populations generally increase, owing to an increase in nutrient levels (Fig. 4-11). As the filling con-

FIGURE 4–11

A eutrophic lake. Note the vegetation invading from the left. (*Photo: G. Marquardt*)

tinues, a stage is eventually reached where the thermocline never develops and complete vertical circulation occurs throughout the year. Consequently, the lake will tend to warm above 4°C, and cold-water forms such as trout will be eliminated. The lake at this stage is classified as senescent (Fig. 4-12). It is in this stage that the water levels are so shallow that rooted plants become established in even the deepest part of the lake. Eventually, via normal filling, the senescent stage is passed and the system may become a pond (merely a smaller, shallower area of permanent or temporary open water), or it may develop immediately into a swamp or marsh. (By definition a swamp contains water-loving or hydric trees, whereas a marsh is treeless.) It is to be noted, however, that all ponds and marshes were not, at one time, oligotrophic lakes. In many cases these areas were never deep enough to be considered oligotrophic. In these instances they began as eutrophic or senescent systems or as ponds. It is also to be noted that as any of these areas age and become shallower, environmental changes are of greater consequence, since there is less water volume to compensate for and dilute the influx of these materials. For example, the

FIGURE 4–12

A senescent lake. The root mat and invasion of vegetation are well advanced. (*Photo: G. Marquardt*)

input of large amounts of organic material into an oligotrophic lake would be less harmful than the input of the same type and amount of material into a shallow, senescent lake or pond (see Chapters 7 and 8 for details).

SUGGESTED READINGS

Benton, Allen H., and William E. Werner. 1974. *Field Biology and Ecology*. New York: McGraw-Hill Book Company.

Coker, Robert E. 1954. *Streams, Lakes, Ponds*. Chapel Hill, N.C.: University of North Carolina Press.

Ford, Richard F., and William E. Hazen (eds.). 1972. *Readings in Aquatic Ecology*. Philadelphia: W. B. Saunders Company.

Hynes, H. B. N. 1970. *The Ecology of Running Waters*. Toronto: University of Toronto Press.

Welch, P. S. 1952. *Limnology*. New York: McGraw-Hill Book Company.

QUESTIONS

1 / Explain the food chain.

2 / Distinguish between the biological and physical method of lake zonation.

3 / Explain how light intensity as well as nutrient control productivity in lakes.

4 / Diagram and explain the phosphorus and nitrogen cycles.

5 / Explain how nutrient cycling differs in lakes and streams.

6 / Explain the downhill tendency of nutrients.

7 / Compare and contrast the aging processes in lakes and streams.

8 / Explain why the polarity of water is the key factor in lake stratification.

9 / Explain the chemical and biological implications of prolonged stratification.

10 / Explain the spatial unavailability of nutrients.

CHAPTER 5

groundwater

In addition to surface freshwater systems, another important source of industrial, domestic, and agricultural water is groundwater. Recent data indicate that over 61 billion gallons of groundwater per day is used in the continental United States. This represents 20% of all the water used in this country. Of this amount, 42 billion gallons per day is used in agricultural irrigation, 31 billion gallons is used in rural areas, and 8 billion gallons is used for municipal and industrial purposes. The sole source of this groundwater is that portion of rain or snow that travels below the zone of unsaturated water and enters the groundwater table.

As noted in Chapter 3, precipitation, upon reaching the earth's surface, may follow one of three major pathways. It may (1) enter streams as runoff or surface runoff; (2) evaporate directly from the soil surface or be transpired through photosynthetic processes, and thus return to the vapor state; or (3) move down through the soil. Some of the water moving into and through the soil and subsoil will enter the water table and become groundwater.

groundwater formation

A portion of the water entering the ground will be trapped by rock, soil, and so on, and remain in the upper soil zone (Fig. 5-1) as suspended water. Suspended water is prevented from moving deeper into the soil by the molecular attractions that are exerted on the water by the surrounding soil particles, as well as by the intermolecular attractions exerted by water molecules on each other. Since the spaces between the soil particles in this zone are filled with a mixture of air and water, this portion of the soil is termed the *zone of aeration* and is subdivided into three distinct horizons: the horizon of soil moisture, the intermediate horizon (which may or may not be present), and the capillary fringe.

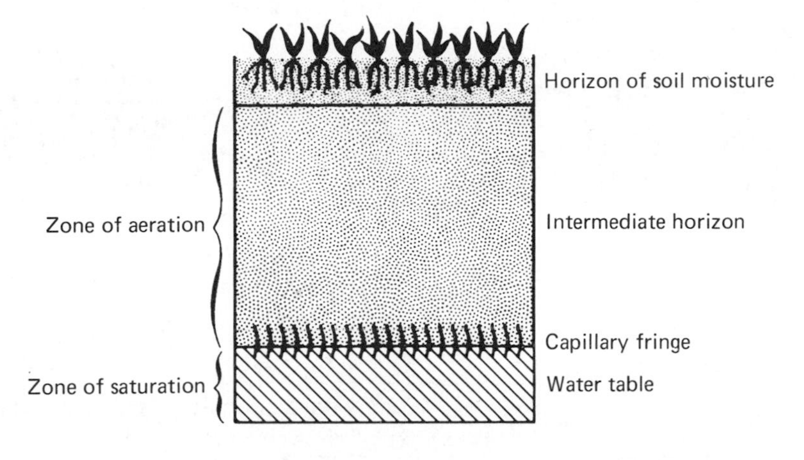

FIGURE 5–1

Soil zonation.

A portion of the water that enters the horizon of soil moisture may be either evaporated or transpired. The remainder passes into the intermediate horizon, where it tends to be held as suspended water by molecular attractions. There is little water movement in the intermediate horizon, except during periods of precipitation, when additional incoming water enters this horizon. In some areas the intermediate horizon is absent, and the horizon of soil moisture lies directly over the third horizon—the *capillary fringe*. Water moves into the capillary fringe (by capillary action) from below.

The zone of saturation lies below the zone of aeration (Fig. 5-1). In this zone there is no trapped air and the openings in the soil are completely saturated with groundwater. The boundary between the zone of saturation is termed the *water table*, which is defined as that point where subsurface water will flow into a well under the force of gravity. The amount, degree of motion, and depth of groundwater is controlled by the structure of the soil and subsoil. Most soils and subsoils are composed of rock and rock fragments that vary in size, density, and compaction. The particles may be small and regular in shape, which results in a close intergranular "fit" with little pore space between the particles. Owing to the low degree of pore space, these subsoils cannot hold much water between the particles and are said to be of *low permeability*. In other words, subsoils of this nature have little ability to hold or transmit fluids. An example of such a subsoil is clay

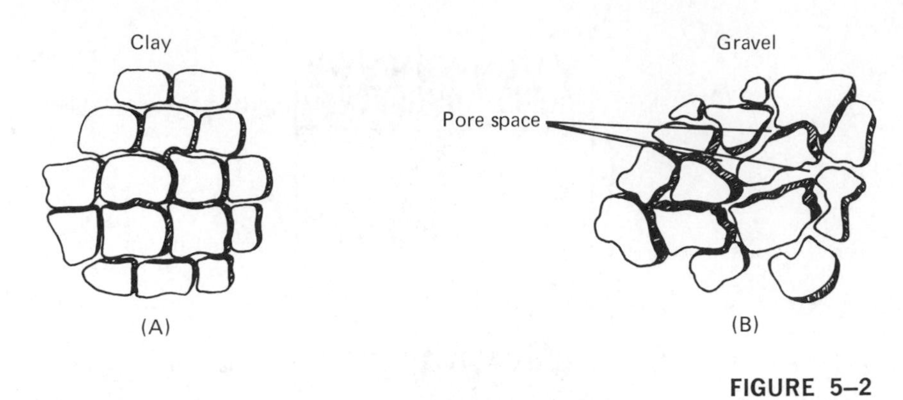

FIGURE 5–2

Typical pore spacing in clay and gravel.

(Fig. 5-2A). Other subsoils may consist of large, irregularly shaped particles that fit together poorly. This type of structure results in many small, interconnecting pore spaces into which water may flow and move. Subsoil of this type has a high ability to hold and transmit fluids and is said to be *permeable*. Subsoils of gravel or sand (Fig. 5-2B) are highly permeable materials. A recognizable portion of the subsoil that is permeable and through which water moves is termed an *aquifer*.

The zone of saturation is formed by water moving through the subsoil under the influence of gravity and eventually encountering a barrier that prevents further water movement. The barrier may be the result of the water encountering materials of low permeability, a dense, compact mass such as igneous rock, or some other barrier impervious to water movement. Water that encounters such a barrier, termed an *aquatard,* tends to completely fill the pore spaces in the aquifer. When this occurs, the pore spaces become saturated and a zone of saturation is formed. The upper surface of this zone, the water table, therefore represents the upper limit to the usable groundwater.

In most cases the water table is located only a few meters below the surface. It may, however, be at the surface, as at the shores of rivers, streams, or some lakes. In other cases it may be at great depths, as in desert areas. Thus the quantity of groundwater will vary geographically in relation to climate and soil conditions. For example, along the Coastal Plain of the United States (a belt of low-lying land along the Atlantic and Gulf Coast), large quantities of groundwater are found in sand, gravel, limestone, and sandstone deposits and provide extensive reservoirs of water. Throughout the Great Plains and the Western and Mountain States, groundwater is much less abundant.

groundwater movement

Below the water table, in the zone of saturation, groundwater tends to flow through the interconnected pore spaces by a process termed *percolation.* In response to the force of gravity groundwater moves from areas of high water table to areas of low water table. Recharge of groundwater occurs by means of the percolation of rain and melt waters into the soil. Because of the variation in both precipitation and permeability, the percolation rates are not uniform, and this leads to an uneven water table which will tend to be high in areas of high rainfall (or high permeability) and low in areas of low rainfall (or low permeability). This uneven distribution causes groundwater to be in constant motion as it flows under the influence of gravity, from areas of high-water-table level (high hydraulic head) to areas of a lower hydraulic head (low-water-table level). In general, groundwater tends to flow toward surface streams and ponds. The rate of groundwater flow is governed by the permeability of the aquifer and the hydraulic head within the system. The relationship between permeability and hydraulic head has been formulated into *Darcy's Law,* which states that the velocity of groundwater is equal to the hydraulic gradient (the differences in water pressure between two points) times the coefficient of permeability. This coefficient represents the degree to which the substrate will permit the flow of water. Darcy's Law is summarized by the equation

$$\frac{Q}{A} = K \times \frac{h}{l}$$

where Q represents the volume of groundwater transmitted per unit time and A is the cross section of the area under consideration. The vertical drop of water from one height to another is represented by h and the distance traveled by the water is l. The permeability of the material through which the water flows is accounted for by multiplying the right side of the equation by a proportionality factor (K). K, therefore, represents the permeability of the soil or subsoil or, in other words, how easily this material transmits water.

From this equation the velocity of flow can be determined or, if the velocity is known, the volume of flow (hydraulic conductivity) through the sediment can be calculated. Velocities and direction of groundwater flow are generally determined by introducing dye into recharge wells and monitoring adjacent test wells until the dye travels to these sites. An alternative method, used when dissolved materials interfere with the detection of dye, is adding sodium chloride to the recharge well and monitoring the test well until the chloride ions are detected. These methods are discussed in Appendix III.

Groundwater movements are slow in comparison with surface waters. Movement in deep aquifers ranges from less than 0.5 centimeter (cm) per day to approximately 100 cm/day. In the majority of aquifers, groundwater moves at rates of only a few centimeters per day, whereas in very permeable soils and subsoils near the surface, velocities may be as high as 15 cm/day.

groundwater quality

Groundwater is always free of suspended materials because the sediments that comprise the aquifer act as filters and effectively remove all particulate matter. These sediments do not, however, remove many of the dissolved materials from the water; since water is considered to be the universal solvent, it can be expected to have a high concentration of dissolved materials. For example, as noted in Chapter 3, particles released as smoke from industrial processes may serve as nuclei around which water vapor can condense. When this occurs, these particles will dissolve in the newly condensed water and will be returned to earth in their dissolved form. Upon returning to earth, they may be carried into surface-water bodies as runoff or into the aquifer with the groundwater. So, in this case, the quality of the groundwater is dependent on the quality of the air through which the water vapor traveled. Another type of material that may enter groundwater is comprised of inorganic plant nutrients that enter from cesspool and septic-tank recharge or as a result of the fertilization of lawns and agricultural lands (other sources of groundwater contamination are discussed in Chapter 8).

Once materials dissolved in the water percolate into the soil, they leave the euphotic zone. Consequently, in aquifers there is a total absence of vegetation and, therefore, no food chain as such. Certain types of bacteria may be able to survive in these areas by utilizing organic materials that are carried in from the surface and by converting these materials into their inorganic forms (o-PO_4 and NO_3), which will increase the concentrations of these materials. In the absence of plants, however, these materials cannot be utilized. Thus, the aquifer will serve as a reservoir of spatially unavailable plant nutrients and other dissolved materials.

use of groundwater

The major method of groundwater utilization is by means of wells. In early times, a well hole was dug below the level of the water table and the subsoil was retained by means of wood or ceramic materials. This permitted the water to flow into the well from the surrounding saturated zone. Presently, most wells are constructed by driving or drilling a well pipe into the saturated zone. The well pipe may be of variable diameter and have at its lower end a well point, which consists of a reinforced point and a screen which allows water through but prevents entry into the well of the fine sediment that composes the aquifer. The screen openings may vary in size, depending upon the type of subsurface materials encountered, and may be formed of actual screening or, more frequently, of stainless steel wire wrapped over a noncorrosive metal frame.

The well pipe fills with water to the level of the water table, and water is generally pumped from this level to the surface by various means. Lifting water from the well lowers the level in the well and produces a *cone of depression,* a conical depression in the water table in the immediate vicinity of the well. Large-volume irrigation or industrial wells can produce a very wide and steep cone of depression, which may significantly lower the water table over a large area. Such a lowering of the water table is referred to as *drawdown* (Fig. 5-3).

In many urbanized areas, increased groundwater pumpage for municipal or industrial purposes, coupled with decreases in natural recharge to the aquifer (due to paving of roads and construction of housing developments, etc.), has resulted in drawdowns of many tens of feet over wide areas (see Chapter 8). Withdrawal of more water than is recharged by precipitation results in groundwater depletion.

Along seacoasts drawdown of the water table may permit the encroachment of salty water into local aquifers, making them unfit for use. This is termed *saltwater intrusion.* In areas where aquifers come in contact with ocean water, the fresh and salty groundwater do not mix readily but form a boundary or *zone of mixing.* The boundary is not vertical but slopes gently, with the more dense salt water forming a wedge beneath the fresh water. The position of the boundary is the result of the hydraulic head exerted by the fresh water. Lowering of the water table reduces the hydraulic head and permits the landward migration of the boundary or interface; raising of the water table causes the seaward migration of the boundary. Groundwater depletion caused by excessive drawdown in coastal areas lowers hydraulic pressure and results in saltwater intrusion into local aquifers, which forces the abandonment of wells in these areas.

FIGURE 5–3

Irrigation of agricultural fields may produce a cone of depression and lower the water table. (*Photo: G. Marquardt*)

SUGGESTED READINGS

Davis, Stanley N., and Roger J. M. DeWiest. 1966. *Hydrogeology.* New York: John Wiley & Sons, Inc.

Ettlinger, Ken W. 1974. *Long Island Groundwater: An Environmental View.* New York: Moraine Audubon Society.

Gilluly, James, Aaron C. Waters, and A. O. Woodford. 1968. *Principles of Geology.* San Francisco: W. H. Freeman and Company.

QUESTIONS

1 / Trace the passage of water into the soil and subsoil.

2 / Distinguish between the zone of aeration and the zone of saturation.

3 / Define the water table.

4 / Explain how water tends to accumulate in an aquifer.

5 / Explain Darcy's Law.

CHAPTER 6

the marine environment

Although the marine environment encompasses such chemically, physically, and biologically diverse systems as estuaries, coastal waters, and the deep ocean, all have several factors in common. The factors serving to relate these systems are their distinctive salt content (salinity), sunlight and temperature relationships, the mode of nutrient cycling, and the mechanisms that form currents.

general characteristics

salinity

Salinity is defined as the total amount of dissolved material present in 1 kg of water, assuming that all the carbonates have been converted to oxides, the bromides and iodides replaced by chlorides, and the organic substances oxidized. It is to be stressed that salinity is a measure of the total dissolved material and, therefore, suspended particles do not contribute to the salinity of any given sample. In addition, since salinity is based on the dissolved material per kilogram, and since 1 milliliter (ml) of water weighs 1 g, 1 kg would be equivalent to 1000 ml of water. Thus salinity is actually a measure of the amount of dissolved material (generally in ionic form) per thousand parts of water (1000 ml = 1000 g = 1 kg). Salinity is expressed in parts per thousand (‰) and customarily abbreviated as S‰.

The characteristic salinity of the marine environment originates from the weathering of terrestrial (terrigenous) sediment and from *juvenile water* (water never before in the liquid state) that enters into various regions in the deep ocean. Terrestrial weathering processes are responsible for bringing both dissolved and suspended materials into the marine environment (insoluble suspended sediments are discussed on p. 89). As rainwater falls through the air it comes in contact with and dissolves atmospheric carbon dioxide. On falling to earth the terrestrial material that the rainwater tends to encounter more than any other is feldspar. The basic molecular structures of the feldspars are aluminum, silicon, and oxygen combined, in various ratios with positively charged potassium, calcium, or sodium ions. The carbon dioxide–rich rainwater readily forms hydrogen bonds (Chapter 1) with the alkali ions (Na^+, K^+, Ca^{+2}) contained in the various forms of feldspar. This results in a solution of sodium, potassium, and/or calcium ions, together with bicarbonate ions (HCO_3^-), in addition to the undissolved, suspended hydrated silica (SiO_2) molecules and residual feldspars which have been converted (by solute–solvent interactions) into the insoluble clay kaolinite. This material will be carried downstream and will encounter additional feldspars, which react in a similar manner and further increase the dissolved ion concentration of the stream. If the drainage travels for great distances, this water will contain relatively high concentrations of dissolved ions upon reaching and entering the marine environment. It is to be noted that the elements readily dissolved in these reactions are derived from either ionic or polar terrigenous sediment which, as discussed in Chapter 1, would be expected to react readily with the polar water molecules. The majority of elements contributing to the salinity of marine systems originate from terrestrial sources. The only elements that are believed not to have a strictly terrestrial origin are chlorine, bromine, and iodine ions, which have their origin as juvenile water.

Juvenile water enters the ocean in regions where the sea floor appears to be spreading (rift areas). In these areas the juvenile water is accompanied by molten basalt. These materials flow upward into the rifts and ultimately come into contact with the seawater that overlies the rifts. Recent analysis demonstrates that the ions dissolved in this juvenile water are precisely those ions (Cl^-, Br^-, I^-) not found in stream water as a result of terrestrial weathering and erosional processes. Furthermore, the chlorine concentration (in the form of chloride ions) in juvenile water is almost identical to the chloride ion concentration of seawater (approximately 18.9 g of Cl^-/kg). Juvenile water, as well as the water brought in by stream flow, is efficiently mixed and circulated by the currents and winds. Consequently, large concentrations of these ions do not build up in any given area but, rather, are uniformly mixed and distributed throughout the marine systems of the world. The mixing process is so efficient that the elements present in seawater are found in uniform ratios throughout the entire world. Seawater, therefore, is said to exhibit a *constancy of composition,* meaning that all the elements in uncontaminated seawater are present in constant, uniform ratios.

Although it has been demonstrated that the ratios of the various elements in seawater do not vary, analysis of seawater samples taken from different geographical areas will show that the actual salinity of the water will be quite variable, especially in the coastal waters. This variation in actual salinities, however, does not contradict the constancy of composition of seawater. For example, the two most abundant elements found in seawater are the chloride ion, comprising approximately 55% of the total dissolved ion concentration of any seawater sample, and the sodium ion, which accounts for approximately 30% of the total. Comparing salinities of seawater samples taken from the deep areas of the North Atlantic with a typical inshore area such as Long Island Sound would generally show the oceanic waters to have a salinity of approximately 36‰, and the coastal areas, as a result of such factors as river dilution, to have a salinity of approximately 26‰. If the actual chloride ion concentration of the two samples were to be determined (assuming that chlorine comprises 50% of the total dissolved ion concentration), it would be found that the Atlantic water had a chloride-ion concentration (chlorinity: abbreviated Cl‰) of 18‰. In the example given, the salinity was determined to be 36‰ and, since the chloride-ion concentration comprises 50% of the total dissolved ion concentration, the chlorinity would be exactly half the salinity (Cl‰ = 18). Similarly, Long Island Sound water, with a salinity of 26‰, would have a chlorinity of exactly 50% of the salinity (Cl‰ = 13). Thus in each case, regardless of the actual salinity, the ratio of the dissolved constituents would be in an absolute invariable ratio or percentage. (Methods of salinity and chlorinity analysis are given in Appendix III.)

density

Obviously, the presence of dissolved material in seawater causes a given amount of this water to be "heavier" (denser) than an equivalent amount of fresh water. If 1 liter of the Long Island Sound water (S‰ = 26), 1 liter of North Atlantic water (S‰ = 32), and 1 liter of lake water were obtained and compared on the basis of weight, the Atlantic water would be the heaviest, the Long Island Sound water next in weight, and the fresh lake water the lightest, even though equal volumes were weighed. This is due to the variation in density between the three samples. As discussed in Chapter 1, density is the relationship of mass to volume ($D = M/V$). Consequently, although the volumes are equal, the masses vary, causing density differences between the various water masses. Thus when marine water encounters fresh water or water of a lower salinity, the waters will tend to layer on the basis of salinity and density variations, with the most saline water tending to sink below the less dense (and less saline) waters. It is for this reason that saltwater can be located considerable distances upstream in rivers as bottom or subsurface water. (See Chapter 11 for methods of vertically sampling a water column.)

As discussed in Chapters 1 and 2, the temperature of a given water mass also influences its density. This is due to the tendency of colder water to have a lower molecular motion and, thus, to form a greater number of hydrogen bonds, thereby decreasing the volume and increasing the density. Seawater, however, unlike fresh water, does not reach its point of maximum density at 4°C. Rather, the density of the liquid water will continue to increase as the temperature decreases. Since ice crystals exclude foreign particles (molecules, atoms, and ions other than water) from their structure, the concentration of dissolved materials in liquid water will increase as ice formation increases. This is due to the fact that, as discussed in Chapter 1, the majority of excluded particles are forced out of the ice crystals and into the underlying, unfrozen water. As the temperature of a marine system decreases, the salinity of the unfrozen water will continue to increase. In addition, there is a direct relation between freezing point and salinity, since, as the dissolved ion concentration increases, the freezing point of the unfrozen water will decrease. In light of the above, therefore, the salinity of marine water at and below the normal freezing point is directly influenced by the temperature of the water column. In addition, the density of marine systems is directly related to both temperature and salinity relationships.

In the very deep ocean, pressure as well as temperature and salinity directly affect the density of the water. In these instances the great pressures generated by the tremendous weight of the water overlying a given deep area will tend to force the water molecules into a much closer proximity than they would normally assume. In these cases the volume occupied by a given number of molecules (and their dissolved ions) is decreased while the mass is increased, thus increasing the density. In some instances, such as in the deep waters of the Mindano Trench adjacent to the Philippine Islands, a temperature rise is noted that is due to the high pressures forcing the water molecules closer together. This phenomenon is termed an *adiabatic temperature change.*

The interaction of temperature, salinity, pressure, and density on a particular water column may serve to generate one of two basic types of currents, termed *pressure gradient currents* (PGCs). A hypothetical model for the first type of pressure gradient current (PGC-I) is illustrated in Fig. 6-1A and B. In this example, keeping variables to a minimum, it can be assumed that the temperature and salinity are

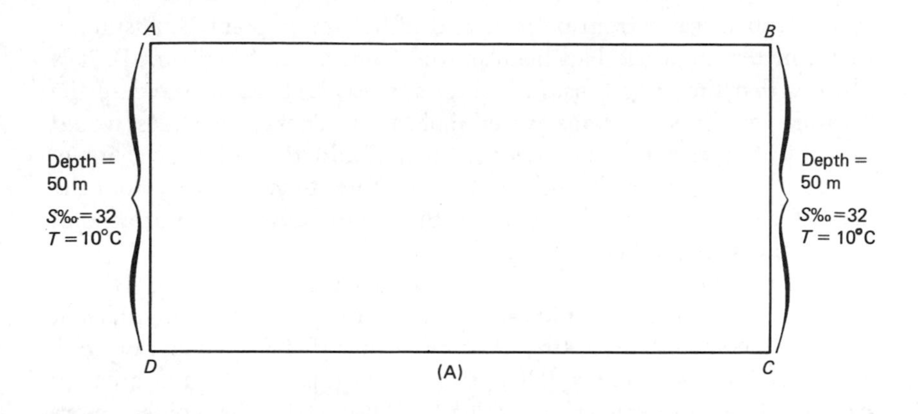

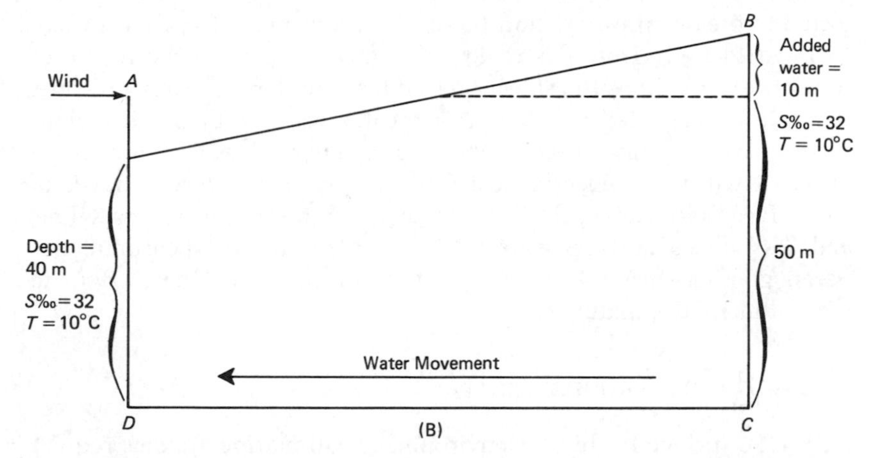

FIGURE 6–1

In (**A**) the water overlying points *D* and *C* is uniform (50 m). In both (**A**) and (**B**) the water is isohaline (S‰ = 32) and homothermous (T = 10°C). In (**B**) the water is moved from point *A* to point *B*. This places additional water over point *C*. The mass of this additional water increases the pressure on the water molecules at point *C* and causes them to move toward point *D*—an area of lower pressure—as a pressure gradient current type I.

uniform (homothermous and isohaline) throughout the system and that the water depth is constant. In certain areas of the world (Cape Hatteras, for example) winds tend to blow predominantly in one direction. In these instances (Fig. 6-1A) this type of wind action would tend to move water from point A and pile it up at point B. Assuming that 5 meters of water had been moved from point A to point B, it is obvious that the added mass of water at this point would increase the pressure on the subsurface water molecules. These molecules would be subjected to a greater pressure than would the subsurface molecules underlying point A and would, therefore, tend to move from the area of higher pressure (point B) to the area of lower pressure (point A) as a pressure gradient current Type I.

The second type of pressure gradient currents (PGC-II) are formed in response to salinity and temperature differences in adjacent areas of the sea (see Fig. 6-2). Assuming that there is a mass of cold, highly saline water ($T = 0°C$; $S‰ = 38$) adjacent to an area of warmer, less saline water ($T = +5°C$; $S‰ = 34$), the colder, more saline water would be at a greater density than the warmer, less saline water. Thus the water molecules in this colder region would be subjected to greater pressure, and the entire water mass, from top to bottom, would move from this region of higher pressure to the region of lower pressure. Since the subsurface water is under a greater pressure, it would be expected to move at a greater velocity than the surface waters. As this dense water mass encounters progressively less dense water, it will sink beneath the less dense water and form a layer of subsurface cold, saline, dense water beneath the warmer, less saline, and less dense surface waters. By precise salinity measurements, oceanographers are able to trace water currents from their origin to their ultimate destination.

nutrient relationships

Like surface freshwater environments, all marine systems require sunlight as an energy source and a sufficient supply of inorganic nitrate and phosphate in order for marine plants (primarily phytoplankton) to carry on photosynthesis and produce the food necessary for the other organisms in the marine food chain. Marine systems are, like freshwater systems, customarily divided into an upper euphotic zone of sunlight penetration and photosynthesis, a dysphotic zone, and the benthos (photosynthetic–respiratory relationships and zonations within the water column are discussed in Chapter 4).

In the deep ocean there is an effective temperature–density barrier similar to those that develop in oligotrophic lakes during stagnation periods. This barrier is even stronger in marine systems, however, since the salinity variations tend to reinforce the purely temperature-induced stratifications that develop in freshwater systems. In addition, the deep ocean currents tend to move not only the deep waters but

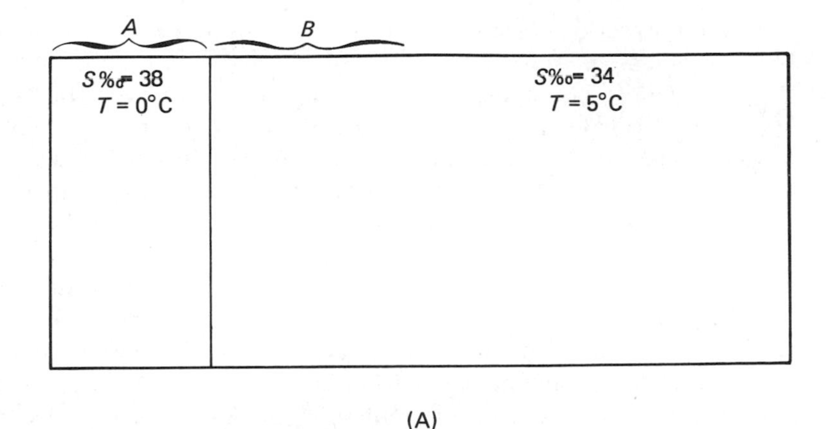

(A)

A
B

(B)

FIGURE 6–2

In (**A**) a mass of highly saline, cold water is generated at point *A*. Since this water is under a greater pressure than the adjacent, warmer, less saline water (point *B*), this water will tend to move from top to bottom into the area of lower pressure. Since the water generated at point *A* is denser, it will sink beneath the less dense water (**B**).

any nutrients dissolved therein. Thus in the deep ocean the nutrients are not only spatially unavailable when they leave the euphotic zone, but are also carried far from their area of original input. This causes vast nutrient redistributions in the deep ocean. The net flow of nutrients in the world's oceans tends to be away from the temperate and tropical regions and toward the poles.

In the deep ocean, nutrients in the proper form tend to be available only seasonally, which results in a cyclical abundance of phytoplankton. The factors that cause this seasonal fluctuation are the amount of incident sunlight, the concentration of nutrients that are spatially available, the depth of the surface-water layer that is mixed by the wind, and the abundance of primary consumers. The amount of sunlight available for photosynthesis depends upon the season and the geographic latitude. In the polar regions the seasonal variation in sunlight is extreme; consequently, the phytoplankton are able to grow and reproduce for only a short period of time. In the tropics, on the other hand, there is always sufficient sunlight, since seasonal changes are minimal.

Nutrients are removed from the water column by the phytoplankton and converted to organic phosphorus and nitrogen. As phytoplankton growth and reproduction progress, these nutrients become chemically unavailable for use by other phytoplankton. If the nutrients are not replaced from the dysphotic zone, where nutrients accumulate, or from the regeneration zone, where bacterial action converts the organic phosphates and nitrates into their inorganic forms, the nutrients will become depleted and will limit further phytoplankton growth. Replenishment occurs primarily through wind mixing.

The amount of water that can be mixed depends on the intensity of the wind as well as the density differences of the water column. Since both density and wind intensity vary seasonally, the amount of water mixing will also vary seasonally. Intense mixing tends to bring spatially unavailable nutrients from the dysphotic zone up into the euphotic zone to be used by phytoplankton.

As the phytoplankton increases seasonally, the food supply of the primary consumers is increased. This leads to an increased consumer population and an increase in the predation on the phytoplankton. As grazing continues, phytoplankton populations decrease, and the nutrients are tied up in the animal portion of the food chain, resulting in a chemical unavailability of the nutrients necessary to sustain high phytoplankton levels. The yearly cycle for the north temperate region, involving nutrient levels, phytoplankton, and primary consumer levels, is given in Fig. 6-3.

Because these factors vary geographically as well as seasonally, the world's oceans can be conveniently divided into three geographic regions on the basis of their nutrient relationships. The tropical oceans are characterized by a continually high degree of sunlight penetration into these waters. Since the insolation (sunlight) is constant, the water

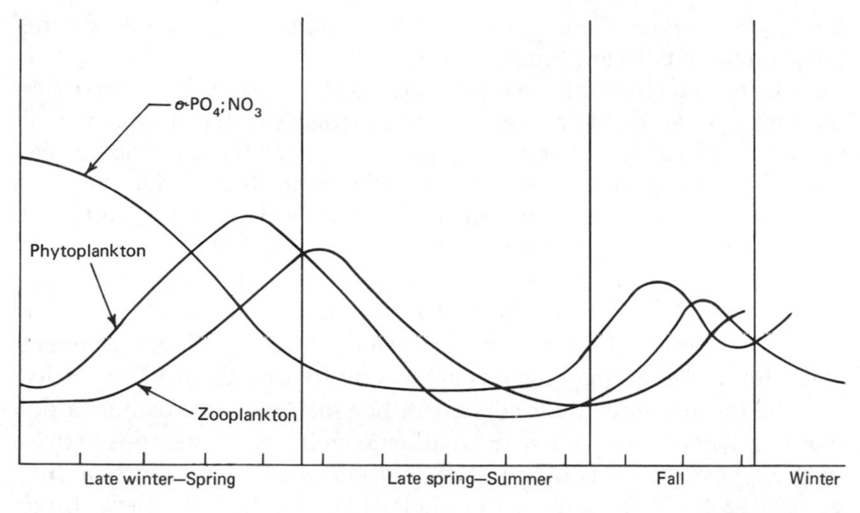

FIGURE 6–3

Yearly nutrient–phytoplankton–consumer curve (north temperate zone).

temperatures at the surface are high, which results in a strong, well-defined thermocline that is resistant to wind mixing. The surface, mixed layer cannot be easily refertilized from the dysphotic zone, causing low nutrient levels throughout the year, which results in low phytoplankton populations.

In the temperate regions there are definite seasonal fluctuations in nutrient levels. Cooling of the surface waters tends to reduce or lower the temperature–density barrier, while winter storms bring nutrient-rich water from the dysphotic zone up into the epilimnion. However, these nutrients are not used immediately since sunlight, in winter, is insufficient to sustain a large phytoplankton population.

In the spring a combination of warmer temperatures and increased sunlight stimulates a phytoplankton bloom. This bloom is soon followed by a zooplankton (primary-consumer) bloom. Active grazing by the zooplankton during this period decreases the numbers of phytoplankton and ties the nutrients up as animal tissue.

During the summer most of the nutrients are in their organic form and are unavailable for the phytoplankton. In addition, the surface waters gain heat, reinforcing the temperature–density barrier. The possibility of effective water mixing is decreased since storms of sufficient intensity generally do not occur during this season. These factors prevent nutrient-rich water in the dysphotic zone from entering the euphotic zone. Consequently, phytoplankton populations during the summer months are generally low.

In the fall the water temperatures begin to cool. This decreases the strength of the temperature–density barrier. The frequency of storms increases in the autumn, which, in combination with a decreased temperature–density barrier, allows nutrients from the dysphotic zone to be brought into the epilimnion. Although the sunlight is decreasing, it is still sufficient to promote a fall plankton bloom.

Nutrient variations, and thus plankton blooms, in the polar regions are similar to those that occur in the temperate oceans. The growing season is shorter, however, while storms are more frequent and intense. The spring bloom is delayed until after the pack ice melts, and the fall bloom is the exception rather than the rule, owing to decreasing daylight and adverse weather conditions. Under the permanent ice cover there is only one bloom per year. This occurs in July or August after the snow cover melts from the ice. At these times sufficient sunlight can penetrate the ice to allow for plankton growth.

In shallow bays and estuaries the nutrients cycle much the same as in shallow rivers and lakes. In these shallow areas the bottom is often in the euphotic zone, there is no spatial unavailability of nutrients, and photosynthesis can occur in the entire water column. In somewhat deeper coastal systems the stronger currents and wind action generally effectively prevent the subsurface waters from stratifying. Consequently, even in systems deep enough to have a dysphotic zone, the nutrients seldom become spatially unavailable. Generally, however, even in shallow systems in temperate and polar regions there are only two significant seasonal plankton blooms each year. In these regions the spring bloom is triggered by the availability of nutrients accumulated over the winter, together with an increase in sunlight during this period. A bloom of zooplankton soon follows, which reduces the phytoplankton population and ties up the nutrients in their organic form over the summer. Thus in summer both phytoplankton and inorganic nutrient levels are low. The autumn bloom, which is generally smaller, occurs because the bacteria have reconverted a portion of the organic phosphorus and nitrogen back to their inorganic form by this time. In shallow tropical regions phytoplankton populations are high throughout the year.

Consumers, however, may feed in shallow estuarine regions and then move out to the deep ocean, where, by excretion or death, the nutrients may be transported from the euphotic zone for long periods of time. In addition, through the action of tides and currents, nutrients will also tend to leave these shallow areas and travel to deeper waters, where they will become spatially unavailable for long periods of time.

Terrestrial input in the form of rainwater runoff, stream flow, and the recomposition of marsh vegetation, however, tend to keep nutrient levels high in these areas.

One of the greatest factors that lowers the productivity of any system is the low availability of either phosphate or nitrate. For example, if a system has a phosphate/nitrate ratio of 1:15, the plants will utilize the total amount of phosphate but only be able to utilize half of the available nitrate. In this case phosphate is said to be the factor that limits further plant growth in the system. In other instances the ratio may be reversed; in other words, there will be an excess of phosphate (e.g., two phosphates per seven nitrates). In this case nitrate would be the limiting factor.

In most marine systems (with the exception of the tropical ocean), it is found that nitrate is generally the limiting factor. This is due to the fact that blue-green algae are absent from all marine systems except the tropical oceans. As discussed previously (Chapter 4), blue-green algae have the ability to convert atmospheric nitrogen gas to nitrate. The absence of this form is correlated with the low availability of nitrate in most of the world's marine systems. The major source of phosphorus is from terrestrial areas, which supply marine systems with sufficient phosphate.

On the basis of the discussion above it is apparent that the determination of nutrient levels in marine systems can yield valuable information. Inorganic nitrate and phosphate levels can be used to predict the onset of plankton blooms and to infer the productivity of a given system. In addition, analysis of nitrite as well as nitrate levels will indicate the ability of a system to sustain phytoplankton blooms, while a comparison of nitrogen/phosphorus ratios (both total and inorganic) will indicate the stability of a given system. These analytical methods are discussed in Appendix III.

marine zones and their characteristics

The marine environment, from the familiar beaches to the deep ocean, can be divided into various zones (Fig. 6-4) on the basis of the degree of seawater inundation and the depth of water. The zone immediately above the high-tide zone, which receives water only in the form of rainfall, spray, or splashing wave water, is termed the *supralittoral zone*. Immediately seaward of this zone is the area alternately covered at high tide and exposed at low. This area is known as the *littoral* or

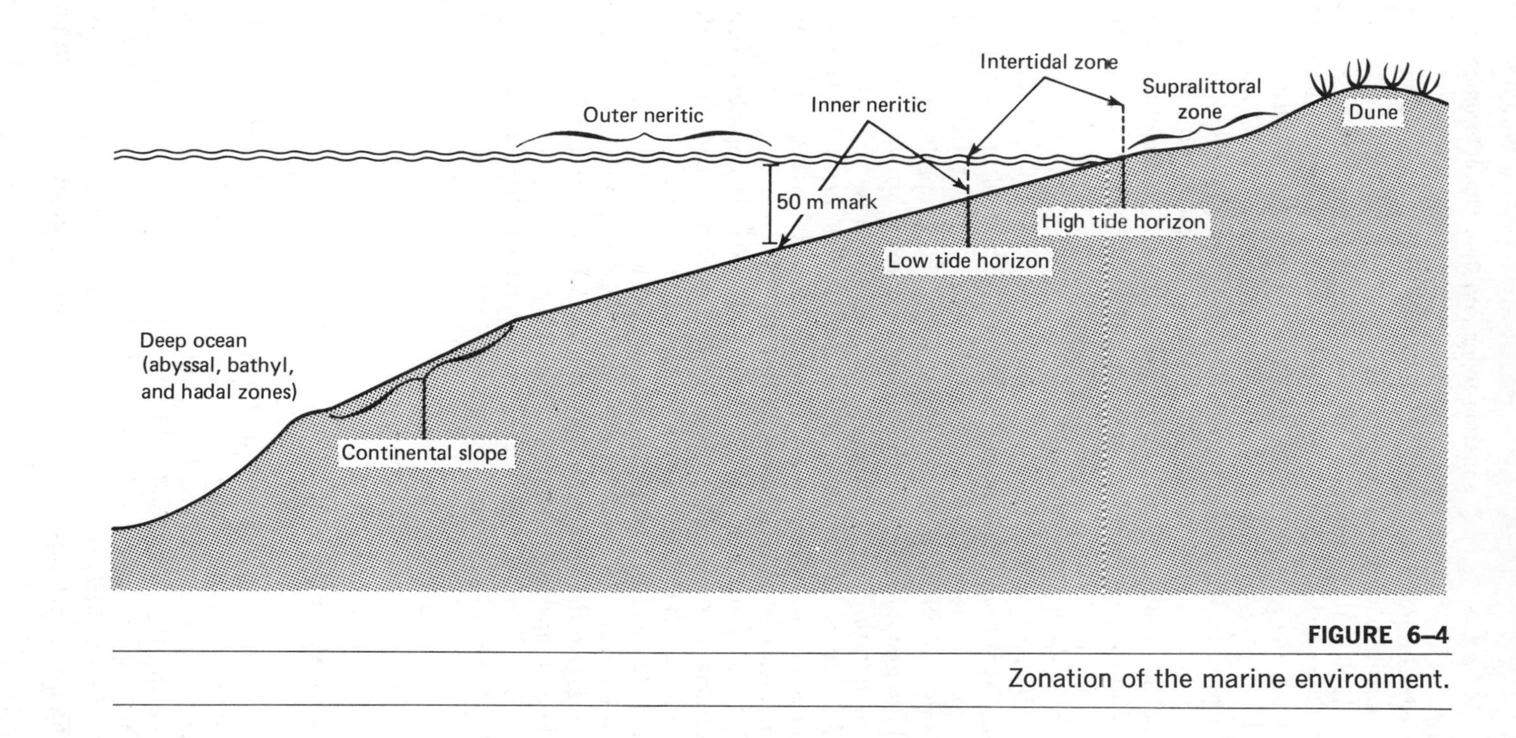

FIGURE 6–4

Zonation of the marine environment.

intertidal zone. Since the animals and plants inhabiting the various portions of the littoral zone are exposed to quite different conditions, depending upon their location in relation to the low-tide mark, the life in this area varies greatly, depending on the amount of time that a given section is exposed to the air. Consequently, this zone is customarily further subdivided into the *high-tide horizon*, which is covered only at high tide and thus subjected to the greatest exposure to air; the *midtide horizon*, receiving intermediate exposure; and the *low-tide horizon*, which receives the least exposure of the atmosphere. Beyond this zone and extending to the continental slope is the *neritic zone*. This zone is also divided into two areas: the *inner neritic*, which extends from the low-tide horizon to the 50-meter (m) mark, and the *outer neritic*, which extends from the 50-m mark (a water depth of 50 m) to the continental slope. The 50-m mark is considered a convenient water depth at which to divide the neritic zone, because it is often the maximum depth to which sunlight can penetrate into the water column. Hence beneath 50 m there would be no effective photosynthesis. From the continental slope to a water depth of 2000 m is the *abyssal zone;* a water depth of greater than 2000 m to a water depth of 6000 meters is termed the *bathyl zone;* and areas under a water depth greater than 6000 m are considered to be in the *hadal zone*.

For the purposes of this text both estuaries and the coastal ocean will be considered to occupy various portions of the inner neritic and intertidal areas. The open ocean will be considered to consist of the outer neritic and the areas farther offshore. Although all these areas are related by their salinity, sunlight relationships, and modes of basic nutrient cycling, they are in other respects (such as salinity variation) quite different and thus must be discussed as separate entities.

estuaries

Estuaries are defined as semienclosed bodies of water having a free connection with the ocean. Estuaries are, therefore, strongly affected by tidal actions. Since they are adjacent to land masses, the marine waters entering estuaries are generally mixed and measurably diluted by fresh water from streams and rivers. Thus any areas in which the marine waters come in contact with and mix with freshwater systems are considered to be estuaries. Common examples would be coastal bays, river mouths, salt marshes, and lagoons (bodies of water behind barrier beaches). Therefore, estuaries can be considered to be transition zones (ecotones) between fresh and truly marine water habitats. Thus they are the initial marine areas to receive the influx of waste materials carried by the river systems.

Since there is a fairly constant influx of fresh water into these relatively shallow areas, there is a wide range of both temperature (since shallow water is more rapidly cooled and heated) and salinity exhibited in estuarine environments. Thus the organisms inhabiting these areas have a wide tolerance to temperature variation (eurythermal organisms) and salinity variation (euryhaline organisms). Despite these rather stressful physical conditions, however, the nutrient availability is extremely high and estuaries are, therefore, the most biologically productive areas encountered.

There are two basic methods of estuarine classification. One method is based on geomorphological relationships and the other on patterns of water circulation and stratification. Geomorphologically estuaries can be divided into five major types: drowned river valleys, estuarine fjords, tectonic estuaries, bar-built estuaries, and river delta estuaries.

Drowned river valleys are the result of a rising sea level, inundating terrestrial regions. They are, thus, formed along coastlines with low and wide coastal plains. Chesapeake Bay is an example of this type of system.

Estuarine fjords are deep U-shaped indentations formed by glacial action. Generally, a narrow sill formed by terminal glacial debris is located at the mouths of these areas. The estuaries found along the coast of British Columbia and Alaska, as well as the Norwegian estuaries, are examples of estuarine fjords.

Tectonic estuaries are formed by movements of the earth's crust. Generally, the movement (faulting) is accompanied by a large influx of fresh water. An example of this type of an estuary is San Francisco Bay.

Bar-built estuaries are generally shallow and often partially exposed at low tide. They are merely an inshore portion of the ocean that has become partially enclosed by chains of offshore islands (barrier islands or barrier beaches) broken at intervals by inlets allowing for an exchange of oceanic and estuarine water. The islands may originate by sediment transported by currents, or they may represent coastal dunes that have become cut off from the mainland by rising sea levels. Long Island's Great South Bay and the sounds (actually estuaries) behind North Carolina's Outer Banks have been formed in such a manner.

River delta estuaries are formed by the deposition and shifting of silt deposits at the mouths of large river systems. Generally this sediment is carried down by the river and is, therefore, terrestrial in origin. The deltas at the mouth of the Mississippi River are examples of such a system.

On the basis of circulation and stratification patterns estuaries may be classified as highly stratified (salt-wedge) estuaries, moderately stratified estuaries, vertically homogenous estuaries, or hypersaline (high-salinity) estuaries. A *highly stratified estuary* results from a strong, dominant flow of river water. In these instances the river water will flow over the denser, saline waters, forming a saltwater wedge underlying the surface fresh water. In large river systems the wedge is capable of extending along the bottom for considerable distances upstream, especially during periods of high tide (see Fig. 6-5). This type of estuary will consist of two distinct layers and will exhibit a salitity profile showing a sharp change in salinity from top to bottom (a *halocline* or salt slope). The mouth of the Hudson River is an example of a salt-wedge estuary, and the halocline is the major reason that marine fish sometimes travel far upstream.

Moderately stratified estuaries are formed when fresh and tidal water flow are more nearly equal in volume and velocity. In these cases the primary mixing is due to turbulence induced by tidal changes. The vertical halocline is less steep and a complex pattern of water layers results from the turbulent conditions. During wet seasons some estuaries will exhibit highly stratified conditions, while, during drier weather, when stream flow is reduced, they will tend to exhibit characteristics similar to a moderately stratified estuary.

Vertically homogeneous estuaries result in cases where the tidal action is dominant. The water will, therefore, tend to be well mixed from top to bottom, and the salinity of the system will be quite high, especially during periods when the tide is rising. Salinity and temperature variations in these instances tend to be horizontal (see Fig. 6-6) rather than vertical and are the result of low-volume stream flow. Bar-built estuaries, such as the Great South Bay of Long Island, are typical vertically homogenous estuaries.

Hypersaline estuaries are formed in regions where the inflow of fresh water is low, the tidal amplitude is small, and the evaporation rate is high. Under these conditions the salinity reaches very high (hypersaline) levels. During very warm seasons a high evaporation rate in these estuaries may lead to salinities higher than those encountered in the deep ocean.

For the purposes of this text estuaries can be divided into two basic life or biotic zones: the flooded estuarine portions, which are always covered by water (those areas below the low-tide horizon), and the mud flats and salt marshes, which are exposed to the atmosphere for various periods of time (the intertidal areas).

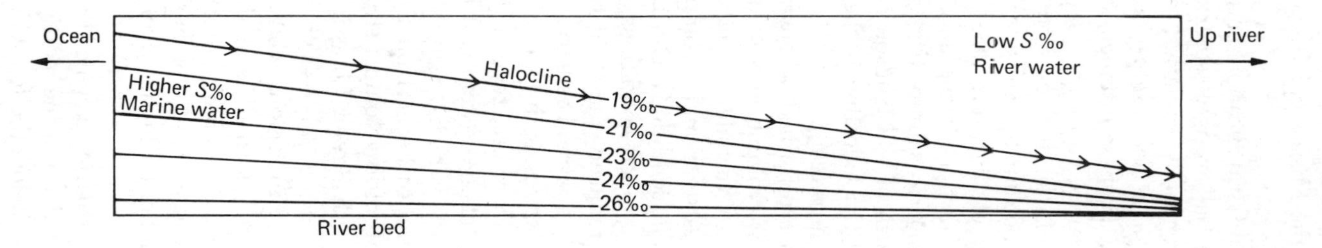

FIGURE 6–5

Salinity profile of a highly stratified estuary. On a rising tide, saline water will enter a river, sink below the less dense river water, and travel upstream as a subsurface salt water wedge. In a system such as this, salinity would vary with depth.

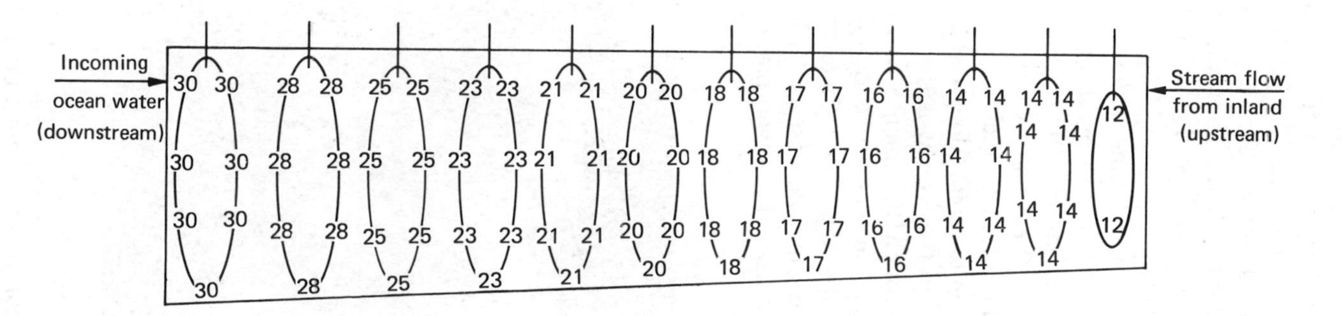

FIGURE 6–6

A vertically homogeneous estuary. Each oval represents a sampling site where water was obtained and analyzed, and the salinity of each sample plotted. Sites of equal salinity are then connected by lines termed isohalines. Note that the salinity is uniform from top to bottom but varies as a source of fresh water input is approached.

Owing to the restricted flow of marine water into estuaries and the concomitant calm conditions encountered in these areas, the sedimentation rate is generally rapid. In these situations the coarser materials tend to settle out near the inlets while the finest silts and muds are carried by the currents to the landward portions of the estuary prior to deposition (see p. 90 for details).

Since the landward portions are generally the shallowest portions of any system, these regions will be the first to break the surface of the water due to this sedimentation. Eventually these landward portions and other protected areas within the estuary will be exposed to the atmosphere for at least a portion of each tidal cycle. Once this occurs the colonization by plants is quite rapid. Along the East Coast, the first plant to become established in these areas is the cord grass—*Spartina alterniflora* (Fig. 6-7). Although this plant "prefers" freshwater environments, it is outcompeted in these areas by other plants and is, therefore, forced into the intertidal zone, where these other plants cannot withstand the tidal flooding. Because of its high tolerance to salinity, S. *alterniflora* colonizes the lower portions of the

FIGURE 6–7

Spartina alterniflora occupies the portion of the marsh that is flooded twice a day. (*Photo: P. Trafas*)

intertidal zone, and the entire area tends to emerge from the water quite rapidly (assuming that the sedimentation rate is equal to or greater than the rise in sea level). This increased elevation occurs because the stems of the cord grass form an effective barrier, and, as the water comes in on a rising tide (with its load of suspended sediments), the stems and leaves of the plants slow the water and cause the sediment to drop out of suspension. Consequently, the marsh tends to become higher, the period of tidal flooding decreases, and the marsh receives less water on rising tides. Thus an effective physical barrier to water inundation is formed, and sediment will begin to be deposited at the seaward edge of the Spartina stands. Eventually, these seaward portions are raised sufficiently to permit water inundation only at high tide, and the S. *alterniflora* moves into and colonizes these newly raised areas of the intertidal zone. It is in this manner that the marsh tends to extend seaward by a combination of physical sediment deposition and S. *alterniflora* growth.

Eventually the rear and calm water portions of the salt marsh that were initially colonized by S. *alterniflora* will have been raised to such a height that they are no longer flooded twice a day. Although these areas provide a suitable habitat for the S. *alterniflora,* they are rapidly invaded by the smaller salt meadow grass, S. *patens,* which can withstand the minimal bimonthly flooding and tends to outcompete the S. *alterniflora* (Fig. 6-8). Consequently, due to competition, S. *alterniflora* is effectively restricted to the portions of the marsh that are flooded frequently and that are unsuitable habitat for the S. *patens.*

On the higher portions of the marsh the S. *patens* is generally associated with two other marsh plants that can withstand this minimal bimonthly flooding—black top grass (*Juncus gerardii*) and spike grass (*Distichlis spicata*). Thus the upper marsh is inhabited by three distinct, dominant species. In many East Coast marshes, *Distichlis spicata* is actually the dominant species, and S. *patens* and *J. gerardii* are said to be associated forms. In disturbed marsh lands (common sights throughout the United States) another plant is found on ground slightly higher in elevation (generally due to landfill operations) and thus generally more landward than the *patens–distichlis–juncus* zone. This is the reed *Phragmites communis,* and its presence indicates disturbed, unnatural conditions (Fig. 6-9). *Phragmites* is extremely harmful because in the fall when its stalks break off they are generally carried out onto the high marsh, where they tend to layer over and kill the indigenous forms. Other organic debris, such as logs, abandoned boats, and eelgrass (*Zostra marina*), also accumulates in these areas and tends to kill the normal vegetation.

The salt marsh can, therefore, be divided into two distinct areas

FIGURE 6–8

Spartina patens (foreground) can withstand flooding only twice a month. Note the reed (*Phragmites communis*) in the background. (*Photo: P. Trafas*)

based on the plant assemblages: the physically lower portion of the marsh, which is flooded twice a day and inhabited by the taller S. *alterniflora;* and the higher portion of the marsh, flooded by bimonthly spring tides consisting of S. *patens–D. spicata–J. gerardii* assemblages in various ratios. On the basis of botanical zonations the relative age of any given marsh can be determined. For example, a young marsh would be composed primarily of S. *alterniflora,* a middle-aged marsh of more-or-less equal amounts of the high and low marsh flora, and an old system would be composed primarily of the high marsh assemblages with only a narrow stand of the S. *alterniflora* occupying the seaward edge. With many states currently attempting to establish legislation for salt-marsh protection based on the location of the various grasses on the marsh, it is necessary that one become familiar with the identification of these indicator species.

The flooded portion of the estuary "ages" in much the same way as does an oligotrophic lake. Initially the bottom is well below the surface and consists primarily of sand or rock substrate. As the water

FIGURE 6–9

Phragmites communis is an indicator of disturbed conditions. (*Photo: P. Trafas*)

conditions become calmer due to the deposition of sand along the seaward portions, the deposition of sediment is accelerated (see p. 94 for details). The coarser, heavier sediment is deposited toward the seaward portions, where the current initially loses velocity. As the current travels landward it continues to lose velocity, and finer and finer sediments are deposited to the rear of the estuary, culminating in the deposition of silts and muds in the calmest portions. Thus a twofold process is continually in operation: the deposition of sediment, which causes a restricted opening to the sea and a shallowing of the bottom. As the opening becomes smaller, the water velocity in the estuary will decrease proportionately, leading to an increased sedimentation rate of, generally, finer and finer sediments. The net result is a closing of the estuary and a decrease in water depth. Eventually a point is reached where sunlight is able to reach the benthic zone (the euphotic zone extends from top to bottom), and rooted submergent plants such as eelgrass (*Z. marina*) and widgeon grass (*Rupia maritima*) become established. Ultimately, if left undisturbed, the system will become so shallow as to provide proper habitat for the *S. alterniflora*, which, as discussed previously, tends to initially extend out into the harbor from its landward stands. Consequently, the end result of these vegetational sequences would be the conversion of the entire estuary to a highly productive salt marsh. This is very similar to the sequences observed in lake aging, as discussed in Chapter 4. A newly formed estuary might be considered similar to an oligotrophic lake, an estuary with well-established, rooted, submergent vegetation and a predominantly S. *alterniflora* marsh similar to a eutrophic system, and an estuary with little open water remaining and the marsh consisting primarily of the high marsh vegetational types (*S. patens, D. spicata, J. gerardii*), similar to a senescent lake.

The major source of nutrients in estuaries is derived from the salt marsh through the saphorphyte food chain. In this type of nutrient cycling the primary consumers do not feed directly on the living plants. Rather, the plants go through their normal growth cycles with minimal grazing from the animal components of these systems. The marsh grass remains largely unconsumed until it dies. At that time it is washed into the open-water portion of the estuary, where it is partially broken down by the decomposers into finely divided particulate matter, termed *detritus*. A portion of the detritus is then fed upon by fiddler crabs (*Uca sp.*) and other primary consumers. From this point the food passes up the food chain in the typical manner. Generally the growing season ends in midfall and the *Spartinas* are washed into the estuary with the first fall storms, enter the decomposition zone, and are reduced to detritus over the winter, thus becoming partially available to the animal components the following spring. Since the detritus is finely divided particulate organic matter, it not only serves as a food source (in its organic form) for many of the animals but is also slowly decomposed into its inorganic components throughout the year. This slow decomposition of detritus into inorganic nitrate and phosphate causes this material to serve as a nutrient reservoir in the sense that it yields plant nutrients during those periods that the phosphate and nitrate, utilized during plankton blooms, are tied up in plant and/or animal tissue. Thus estuarine detritus serves as an additional nutrient source and is a major factor in maintaining continually high productivity rates in these areas. Because of this, the salt marshes play an important role in the nutrient cycles of the estuaries. In addition, the large majority of the nutrients found in the deep ocean originate in the estuaries from the decomposition of the valuable marsh grasses. Consequently, the salt marsh plays an important role in both estuarine and oceanic food cycles.

In reality it is found that the estuarines and the adjacent coastal oceans are the most productive areas in the entire marine environment. There are four primary reasons for this high productivity:

1 / Since they are shallow, more efficient mixing is able to occur via turbulence, wave action, and upwelling caused by winds or offshore currents. These mechanisms serve to prevent spatial unavailability of nutrients by bringing these nutrients into all the water layers.

2 / In addition to the vast amounts of nutrients generated by the salt marshes fringing the open-water portions of the estuaries, additional nutrients are washed into the coastal waters by river and stream flow (see Chapter 3).

3 / In estuaries and the shallow coastal oceans the substrate is generally suitable and the system shallow enough to permit a rich growth of macroscopic plant life.

4 / Since the plants form the base of any food chain, these areas also support a large amount of animal life.

the open ocean

In contrast to the conditions discussed for estuarine and coastal waters, the deep ocean presents an entirely different set of factors. Whereas estuaries comprise only a small portion of the total marine environment, they are responsible for the major input of nutrients to these systems. The open ocean, on the other hand, although it comprises 70% of the earth's surface, is almost entirely dependent on estuarine production to carry out its life processes. In other words, the majority of the nutrients are produced in the esturine environment and transported to the deep ocean, where a portion are used. Most, however, are lost to the euphotic zone and thus become spatially unavailable for long periods of time. In comparison to estuaries and the coastal ocean, the deep ocean is considered to be a virtual biological desert. The primary factor responsible for the low nutrient availability in the deep ocean is its depth and the deep-ocean circulation patterns, both of which interact to keep nutrients spatially unavailable for long periods of time.

In the open ocean there are two basic types of circulation systems: the lesser circulation, which operates from the surface to perhaps 100 m and is prevented from mixing with the deeper waters by the thermocline; and the greater circulation. For the purposes of this text the greater circulation can be considered to encompass all the water below the lesser circulation. For convenience, this water can be divided into the intermediate water mass, the Arctic source water, and the Antarctic source water. The lesser circulation flows from the equator to the poles as a surface-water mass, loses its identity at these polar latitudes, and then reappears as a subsurface water mass at

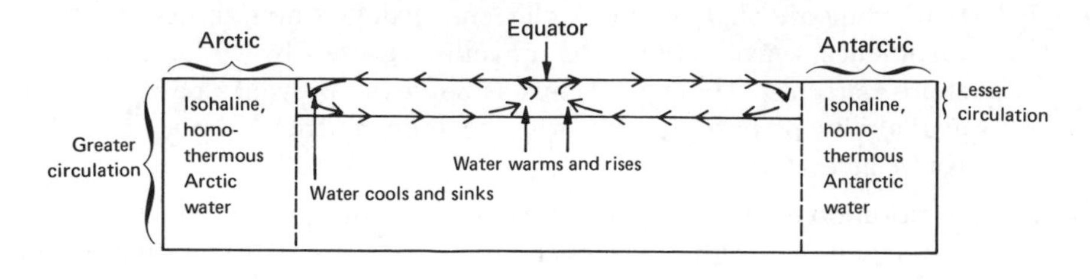

FIGURE 6–10

Lesser circulation.

approximately 100 m flowing toward the equator (see Fig. 6-10). The major factor governing this circulation system is the vast differences in air temperatures between the polar regions and the equatorial regions. At the equator the high temperatures raise the surface waters, causing a decrease in density. These warm, less-dense water masses will then tend to rise above cooler, denser subsurface water and remain at the surface. In the northern hemisphere this warm, surface water then tends to move northward (southward toward the Antarctic in the southern hemisphere). As it moves north, the water encounters cooler and cooler air temperatures and begins to cool down and increase in density. Eventually it reaches the polar regions, cools to the prevailing water temperature, and loses its identity. At this point these waters are extremely cold and, hence, very dense. Because of the increased density of these waters, a pressure gradient current Type II will develop and the cold, dense polar water will begin to move toward the equator (northern water will move south, southern water north). As this water encounters warmer, less-dense subpolar water it will tend to sink beneath this less-dense water. At this point large masses of polar water are moving and sinking. The waters on the exterior portions of this mass (Fig. 6-11) encounter warmer, less-saline subpolar water masses, become diluted, and gain heat and decrease in density. Thus they become separated from the main mass of polar water and travel at a shallower depth (100 m in the case of lesser circulation waters). As this occurs, more and more water is leaving the polar regions (by means of the preceding mechanisms), and this water tends to push the previously formed water toward the equator at varying depths, depending on the temperature–density relationships of these water masses. As this water travels toward the equator it continues to warm, mix with surrounding water and decrease in density. This decrease in density causes the water mass to rise and reach the surface in the equatorial regions, where the process is repeated. Thus the lesser

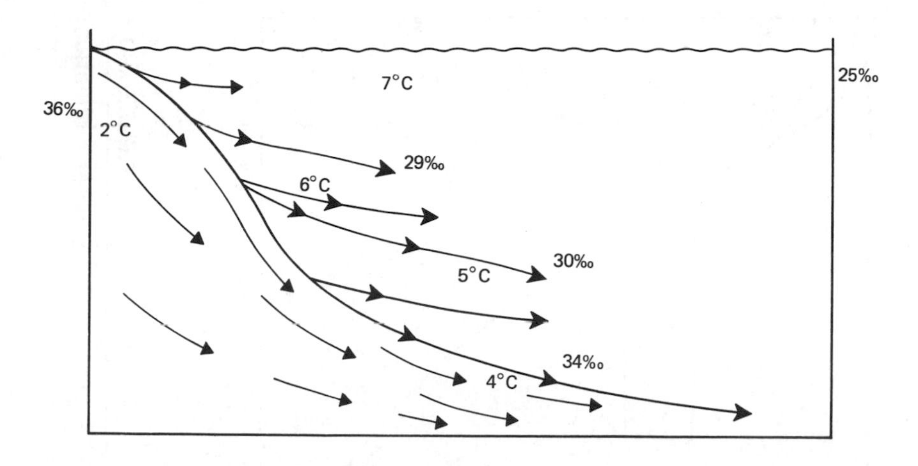

FIGURE 6–11

The path of Arctic source water, depicting the formation of intermediate water masses. As the cold, saline water moves from its source, it encounters warmer, less dense water, and it sinks. The water immediately adjacent to the warmer water mass warms at a more rapid rate, decreases in density, and moves away from the main, dense water mass, thereby forming water masses of "intermediate" temperatures, salinities, and densities.

circulation tends to operate on the basis of warming and cooling of water masses, causing a density difference and a resultant pressure gradient current Type II.

The greater circulation (see Fig. 6-12) operates in much the same manner and is formed by the same type of pressure gradient current. It is to be noted that according to Fig. 6-12, the waters at the Antarctic have a temperature of −1.0°C and a salinity of 36.5‰, whereas the Arctic waters have a temperature of 1°C and a salinity of 35‰. Therefore, the waters of the Antarctic are the densest waters of the world's oceans, and the Arctic waters are the next in density. Consequently, these waters are subjected to higher pressures than the adjacent subpolar waters, and they tend to move from top to bottom as pressure gradient currents (Type II) into the surrounding water masses of lower density. As they move, they encounter this less dense water and sink beneath it. As mentioned above, the water toward the periphery of these sinking masses tends to come in contact with the warmer water as it sinks. This water, therefore, warms and dilutes at a faster rate, levels off, and becomes a portion of the lesser circulation.

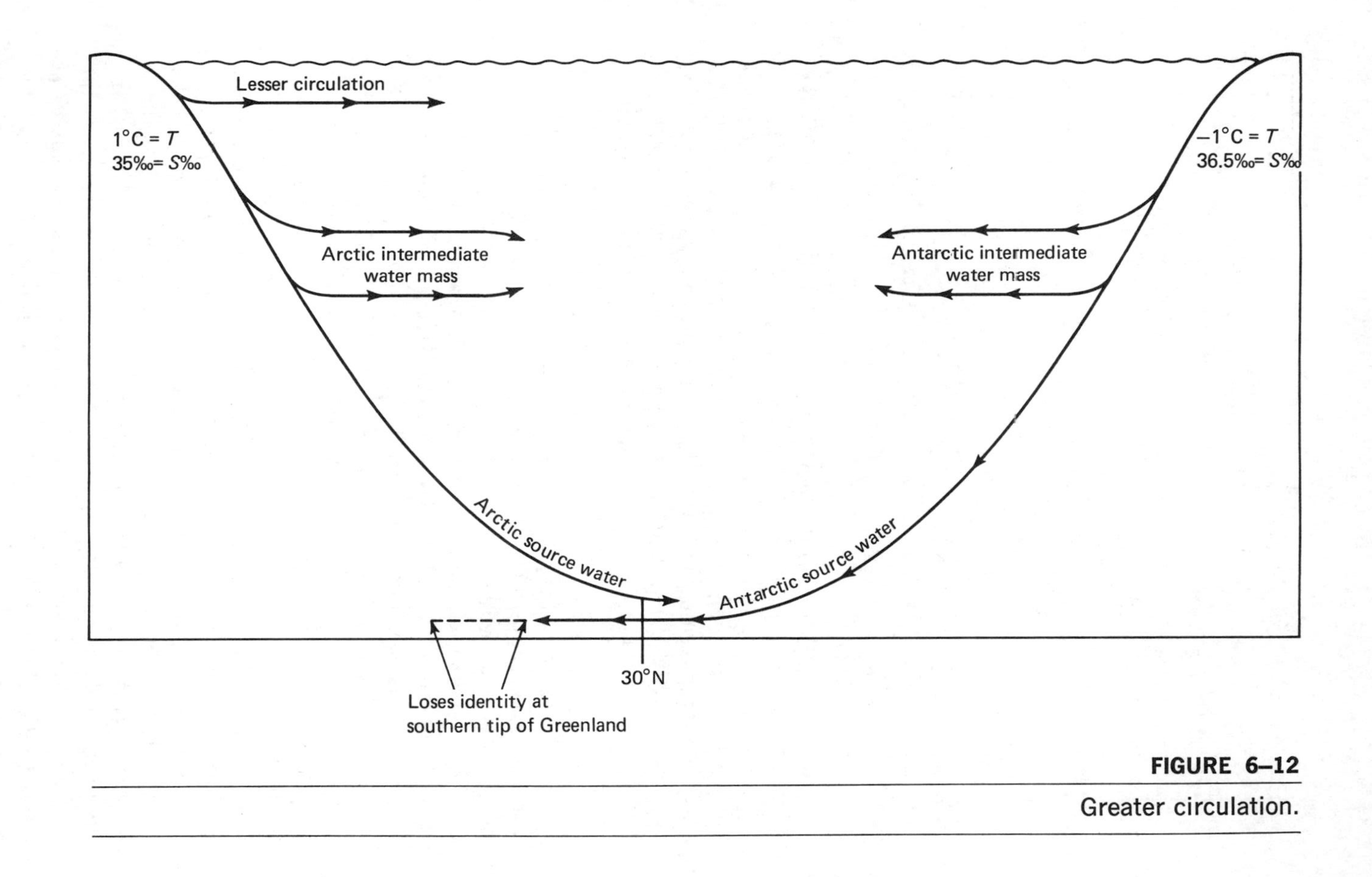

FIGURE 6–12

Greater circulation.

Other portions of the water masses sink to intermediate depths before warming and diluting and, thus, become a portion of the ocean's intermediate water mass. The majority of the water masses from each polar region, however, tend to remain together and sink toward the ocean's bottom.

There are two water masses that travel along or near the ocean bottom. The one originates in the Antarctic, consists of the densest water, flows north as a bottom-water mass, and is termed *Antarctic Source Water*. The other originates in the Arctic, is slightly lower in salinity, higher in temperature, flows south as a deep-water mass, and is termed *Arctic Source Water*.

The Arctic Source Water results from the mixing of highly saline water from the Gulf Stream (see p. 89) with the cold surface waters of the polar and subpolar ocean. This water flows south as a PGC-II, encounters less-dense water masses, and sinks. A portion of this sinking water mass mixes with the surrounding subsurface waters, decreases in density, and thus enters the *Intermediate Water Mass*. The majority of this water retains its identity and moves southward along the ocean's bottom. In the vicinity of the southern tip of Greenland (30° north latitude), it encounters Antarctic Source Water moving northward. Since the Antarctic Source Water is denser, the Arctic Source Water flows above this water mass as it travels southward. Thus the Arctic Source Water flows directly over the floor of the ocean north of 30° latitude. South of this point it flows above the Antarctic Source Water and below the Intermediate Water Mass.

The Antarctic Source Water is, as mentioned above, the densest water in the world's oceans. It is formed by the freezing out of salts, mainly in the Weddell Sea. This cold, highly saline water then flows down the continental slope, mixes with subsurface circumpolar waters, and thereby forms Antarctic Source Water. This water mass moves northward as a PGC-II, encounters less-dense subpolar water, and sinks. As it sinks some of the peripheral water warms, dilutes, and levels off as the Intermediate Water Mass. The majority of the Antarctic Source Water, however, retains its identity and sinks to the ocean's bottom, where it travels southward as a bottom-water mass to the southern tip of Greenland before it loses its identity.

All these water masses mix with the surrounding water as they travel from their source. The salinity and temperature are constantly altered, which results in density changes of the water. Eventually the water will return to the surface, where it will cool or be evaporated or diluted. Ultimately, this surface water will again sink and reenter the

Greater Circulation. Recent studies indicate that Antarctic Source Water, once it sinks, will not return to the euphotic zone for approximately 800 years.

The Greater Circulation serves as a reservoir for much of the world's nutrients. Both organic and inorganic phosphorus and nitrogen from terrestrial regions are carried into the coastal oceans and estuaries via stream flow and rainwater runoff. In these shallow marine systems they enter into the estuarine food chains along with nutrients derived from the salt marshes via the saphrophytic food chains. The organic and inorganic phosphorus and nitrogen are ultimately carried by the estuarine and coastal currents into the deep ocean, where the organic materials enter the regeneration zone immediately and become spatially unavailable, eventually entering the Greater Circulation. The material that happens to be in the proper form (o-PO_4 and NO_3) will be utilized by oceanic plankton, be converted into their organic form, and pass into the animal components of the oceanic food chains. Ultimately these materials will reach the regeneration zone, and thus the Greater Circulation, in the form of excretory material or dead plant and/or animal remains. The Greater Circulation then serves to trap and tie up much of the entire world's nutrients as spatially unavailable phosphate and nitrate. These nutrients will remain dissolved in this water mass until brought to the surface by the slow mixing process discussed above. The current trend of disposing of sewage by ocean outfalls speeds up the loss of phosphorus and nitrogen from terrestrial and freshwater systems, as well as depletes the freshwater supplies of those communities relying on subsurface aquifers as their source of consumptive water (see Chapters 4, 7, and 8).

Other mechanisms for bringing spatially unavailable nutrients to the euphotic zone are illustrated by the circulation pattern of the Gulf Stream. This current originates just north of the equator in the Atlantic Ocean. In this region the prevailing winds are blowing in a northwesterly direction. This wind action sets the waters in motion and they begin moving as a wind drift current, termed the *Equatorial Current,* northwestward up along the coast of South America (see Fig. 6-10). As these waters travel north, they encounter the Antilles Islands, where they are split into two discrete water masses—one traveling to the west of the Antilles, termed the *Caribbean Current,* and one traveling to the east, called the *Antilles Current.* The importance of the Antilles Islands cannot be underestimated, since it is assumed that if they were not present to split the water mass, little, if any, of this water would enter the Gulf of Mexico to form the Gulf Stream (see

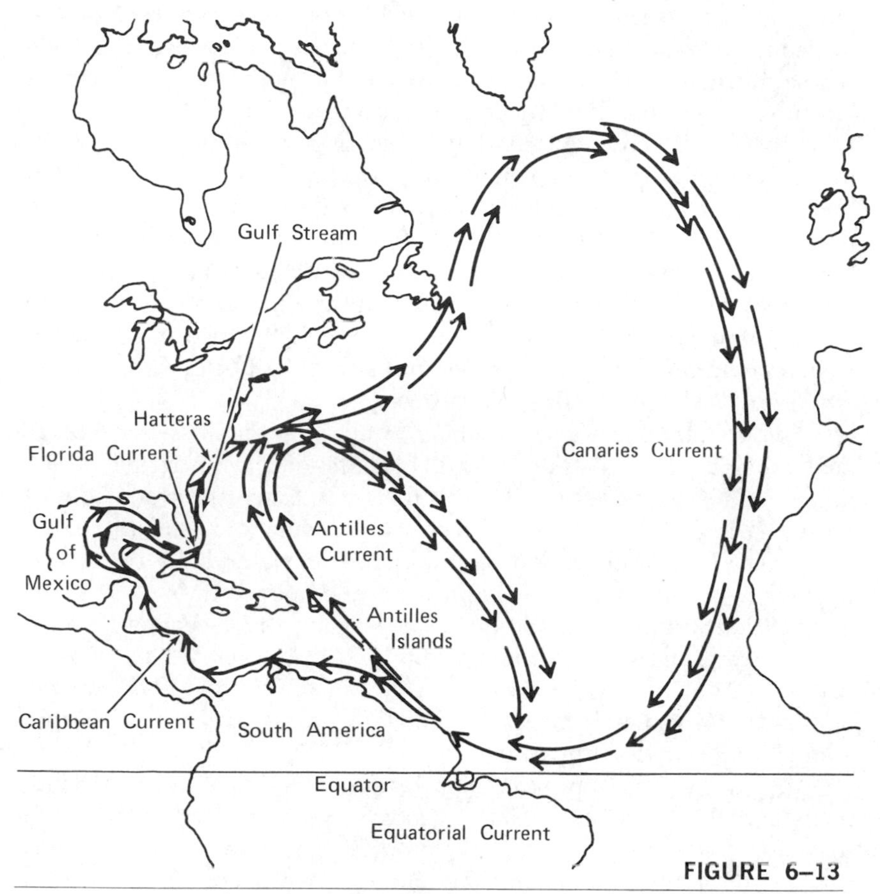

FIGURE 6–13

The major currents comprising the Gulf Stream.

Fig. 6-13). The Antilles Current continues northeastward in a large right-handed arc and will be considered shortly.

The Caribbean Current continues northward between the Antilles and the mainland of South America, and along the coast until it reaches the Yucatan Peninsula. This peninsula has a subsurface extension, called the *Yucatan Sill,* which extends beneath the surface from the tip of the Yucatan Peninsula to approximately 100 miles off the tip of Florida. As the Caribbean Current reaches the Yucatan Sill, the subsurface water is brought to the surface and pushed over the sill into the Gulf of Mexico. This process, whereby the physical presence of the sill forces nutrient-rich subsurface waters into the euphotic zone, maintains a continually high nutrient level in this area. These nutrients sustain a large phytoplankton population and, consequently, a large number of consumer organisms.

Since the winds are constantly blowing water northward from the equatorial regions, the upwelling continually occurs along with the return of large amounts of nutrients to the euphotic zone. Water from the Caribbean Current, therefore, is constantly driven into the Gulf of Mexico, which has only one restricted outlet through the Straits of Florida, located between the tip of Florida and Cuba. The water cannot travel back southward over the sill, since it would encounter additional incoming water from the Caribbean Current and be forced back into the Gulf of Mexico.

This water mixes with incoming water from the Mississippi River and other freshwater systems and, because of the restricted outlet, large volumes of water build up in the Gulf. Since the water is well mixed, it is essentially considered to be isohaline and homothermous. However, as a result of the large volume of water built up in this region, it is subjected to high pressure and moves through the one outlet (the Straits of Florida) as a pressure gradient current Type I. As this water mass moves through the Straits, it is termed the *Florida Current.* As it passes the tip of Florida it is called the *Gulf Stream,* and from this point it moves up along the coast of the southeastern United States as a warm, saline coastal current.

As the waters approach Cape Hatteras, both the Hatteras landmass and the prevailing offshore winds serve to deflect these waters offshore. The winds move large quantities of the surface Gulf waters to the south, thus producing the proper conditions for the second major upwelling that these waters undergo. Referring to Fig. 6-13, it is noted that when wind action moves surface waters from one area to another (point B), the subsurface waters at point B are subjected to a greater pressure and, therefore, move from this area to an area of lower pressure. The winds encountered at Cape Hatteras move the water offshore, which creates a greater pressure on the subsurface waters and causes them to move inshore as a pressure gradient current Type I. As this subsurface water moves inshore it encounters the rising Hatteras shoreline and thus moves upward into the euphotic zone. This combination of shoreward rising water brings large amounts of nutrients into the euphotic zone, where they become spatially available, and large phytoplankton populations are therefore found in this area. The nutrients and phytoplankton are responsible for the productive fisheries located in the Cape Hatteras vicinity.

As the wind and the Hatteras landmass force the Gulf Stream offshore, these waters come under the influence of the *Coriolis Force,* which causes them to travel northeastward as an offshore current in a large right-handed arc.

The Coriolis Force arises due to the effects of a spinning earth on all bodies moving along the earth's surface. For example, the earth is known to rotate on its axis from west to east, and the speeds of this rotation have been calculated for various latitudes of the earth. It is known that at the equator the earth (and all stationary particles on the earth) is moving eastward at a rate of 1670 km/hr (1050 miles/hr), whereas at a latitude equivalent to New Orleans the earth's velocity is only 1450 km/hr (935 miles/hr). Consequently, a stationary particle at the equator would be actually moving eastward at the rotational speed of the earth, 1670 km/hr, while a stationary particle located at New Orleans would be moving eastward at 1450 km/hr (the rotational speed of the earth at that latitude). The same differences in rotational velocities are encountered in the southern hemisphere. If a particle of water is set in motion northward at a velocity of 10 km/hr it would, in reality, be traveling north at a velocity of 10 km and eastward at its original velocity (imparted by the velocity of the earth) of 1650 km/hr. As this particle continues moving northward, it travels into latitudes that are moving more slowly than the velocity of the earth at the equator. The moving water would, therefore, be moving at a greater velocity than the earth at these northerly latitudes. Thus an observer stationed at New Orleans, could he view the entire operation, would observe the following: Prior to any northward movement, the water mass would appear to be keeping pace with the observer. After northward motion was imparted to the water the observer would first notice the water moving directly toward him in a northward track. However, as the water continued moving northward into latitudes of less velocity than those encountered at the equator, the water mass would appear to arc to the observer's east, since the water is now in latitudes that are moving eastward at a lesser velocity than the water mass's initial eastward velocity.

If the procedure were varied and water from the arctic, which is rotating at a lower velocity than either water in the equatorial regions or in the vicinity of New Orleans, is moved southward, the following would be observed. Initially, prior to imparting southward motion the water mass would appear to be keeping pace with the observer stationed in New Orleans. After southward motion of 10 km is imparted, the water will begin traveling south at 10 km/hr and eastward at 1200 km/hr (the speed of the earth at the poles). This water will, consequently, be moving from an area of low velocity into latitudes of higher velocities and will appear to arc to the west as it moves southward into faster-rotating portions of the earth. If observed from a position that is not a portion of the earth (the moon, for example), the water would actually be seen to move in a straight line east-northeast in the first example and south-southwest in the second. The reason that the water appears to arc is that it is being observed by individuals who are on the rotating earth and thus traveling at the earth's velocity at the particular observation points. The Coriolis Force is thus the tendency of water masses to appear to curve to *their* right as they travel in the northern hemisphere and to appear to curve to *their* left as they travel in the southern hemisphere (see Fig. 6-14). The Coriolis Force never initiates water motions but, once water is set into motion, it profoundly influences the direction of water flow.

At Hatteras the Coriolis Effect, with assistance from the Hatteras landmass and the offshore winds, causes the Gulf Stream to travel northeastward as an offshore current in a large right-handed arc. Shortly after passing Hatteras the Gulf Stream brushes past the Antilles Current which (after passing to the east of the Antilles Islands)

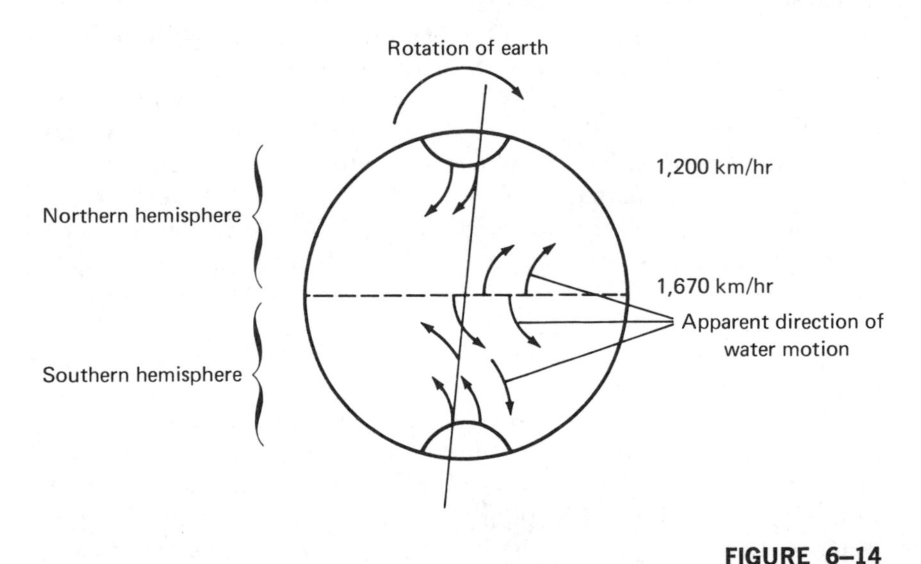

FIGURE 6–14

The coriolis effect.

has continued traveling northerly, under the influence of the winds and the Coriolis Force, in a right-handed arc. The Antilles Current, since it has been exposed to equatorial and subequatorial temperatures during its northward journey, is a warm, highly saline water mass. When it encounters the Gulf Stream the waters of the Antilles Current are under a higher pressure than the Gulf Stream waters are. Consequently, water tends to flow from the Antilles Current into the Gulf Stream as a pressure gradient current Type II and adds to the volume, temperature, and salinity of the Gulf Stream waters. Immediately after passing the Antilles Current (which arcs back toward the equator), the Gulf Stream passes the Sargasso Sea. This body of water has no mechanisms for mixing or upwelling its waters and is, consequently, a warm, highly saline water mass of low nutrients in the euphotic zone. Since it is essentially a nonmoving body of water it has been exposed to tropical temperatures for long periods. Evaporation of water increases the salinity of these surface waters, setting up vertical currents of sinking, dense, surface waters. These vertical currents, therefore, carry nutrients along with the dense water out of the euphotic zone to the bottom of the Sargasso Sea. These subsurface waters are subjected to high amounts of pressure and tend to move out as a pressure gradient current Type II into the adjacent waters of lower pressure. Thus both the Gulf Stream and the Antilles Current receive nutrient-rich subsurface water from the Sargasso Sea.

From this point on, there is little input of water into the Gulf Stream, and it continues traveling northeastward under the Coriolis Force and moderating the temperatures of the North American coastal regions that it passes. It is called the Gulf Stream until it reaches Newfoundland, where it becomes known as the *North Atlantic Current*. This current completes its arc and travels down the west coast of Europe, where it continues to be termed the North Atlantic Current as it travels southward. To the west of North Africa it merges with the *Canaries Current* and is known as the Canaries Current until it passes into the North Equatorial Current, which closes the current circuit. Eventually this water will be blown northwestward by the prevailing winds and will again begin its northward journey.

coastal sediment

As mentioned previously, terrestrial weathering processes bring sediment into streams and rivers, where it is carried into the inshore waters of the coastal oceans and estuaries. The coarser materials are deposited close to shore, while the finer materials, such as the clays and silts, are transported out to deeper waters prior to settling out. Once in the marine environment these terrigenous sediments become mixed with sediments of a marine origin, as well as sediments eroded from coastal areas by wave and wind action. All these coastal sediments form the materials that compose our shorelines.

Although coastal sediments are not strictly a water-management problem, they are considered here since they are formed and transported by water movements and are, thus, directly related to occurrences within the water column. In addition, the sedimentary motion set up by coastal currents causes profound changes in shoreline topography and, especially along the East Coast, is the subject of much controversy from a coastal management viewpoint. Other sediments, such as those of the deep sea, are beyond the scope of this text.

According to Fig. 6-15A and B, a beach consists of all those portions of a coastline in which sediment is actually or potentially in motion out to a water depth of 10 m. A beach is composed of two basic components: a berm and a bar. The *berm* is the above-water portion of the beach and consists of a nearly horizontal terrace of sand brought ashore by wave action. *Bars* are subsurface, elongated mounds of sand that parallel the berm. In general, there are two major types of sediment motion due to water action and relative to the beach face—horizontal and vertical. The vertical motion causes the differences observed in the berm in summer and winter. In summer the berm is generally low and wide, consisting of relatively fine sand, while offshore bars are generally absent. In winter the situation is reversed: the bar is present, and the berm is high, narrow, and composed of coarser sediments. The sediment, therefore, tends to move from the bar to the berm in summer and from the berm to the bar in winter. The reason for this shift is the seasonal change in wave action.

In the winter the wind tends to be stronger and this forms larger waves, which break on the beach face with great frequency and force. These waves tend to rapidly saturate the beach face, causing the water in additional incoming waves to flow seaward back over the beach face after breaking. As these strong winter waves travel back over the beach face, they carry large quantities of sand along with the returning water. This water, with its load of suspended sediment, encounters incoming waves, loses velocity, and deposits the sediment offshore, thus building the winter bar with sediment that was deposited on the berm the preceding summer. In summer the winds and the waves are gentler; consequently, the beach face is seldom saturated with water. These smaller, less-frequent waves travel onshore, remove the finer sediment from the bar, and carry it toward the beach face. When

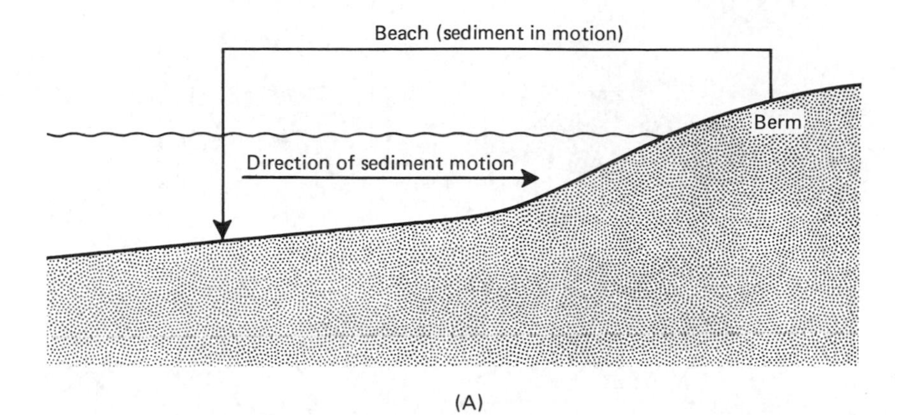

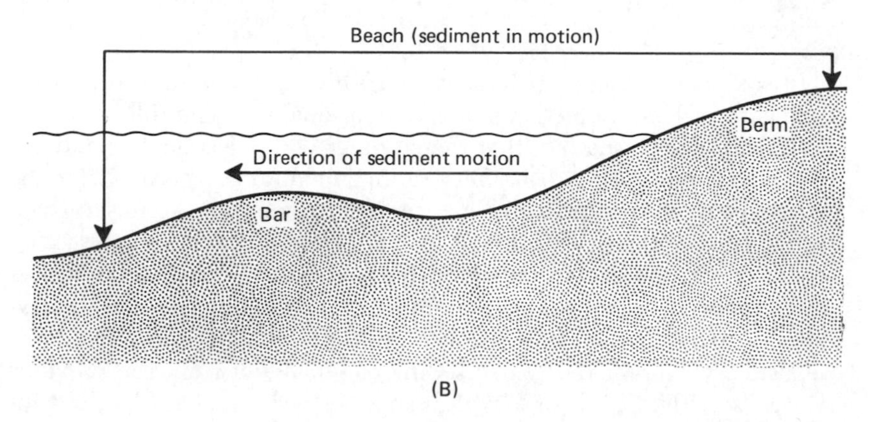

FIGURE 6–15

(**A**) Summer beach and (**B**) winter beach profiles.

these waves break on the unsaturated beach, the water tends to sink into the previously depleted sediment, leaving the newly transported sand (from the bar) in place, which builds the berm.

The sediment comprising a beach, therefore, tends to be transported from the berm to the bar by the strong winter waves, and from the bar to the berm by the gentler summer waves. The sediment profile (see Chapter 11 and Fig. 6-16) of a winter beach is much different than the profile of a summer beach. There is generally such a large difference in winter beach sediment sizes as opposed to summer beach

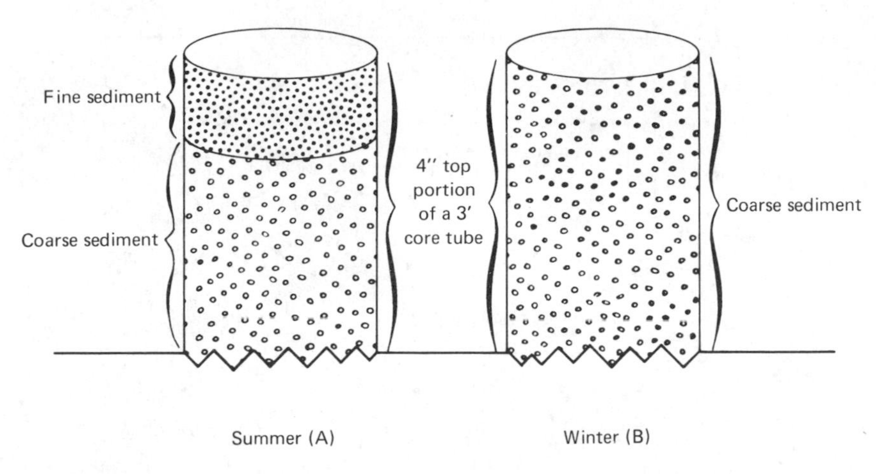

FIGURE 6–16

Sediment profiles of (**A**) summer and (**B**) winter beaches.

sand size that the comparison may often be made visually rather than by means of lengthy methods of sediment analysis. The differences in profile are due to the fact that the summer waves are gentler and are capable of only carrying the finer sediments onto the berm. After the waves break on the berm, the water tends to sink into the unsaturated beach, leaving the new sediment in place. If the beach happens to be saturated, the water will return offshore over the beach face, but at a much reduced velocity. This slow-moving water is incapable of carrying all but the finest sediments offshore. Consequently, the sediment is deposited on top of the coarse sediment remaining from the previous winter (see Fig. 6-15A for a typical summer berm profile). In the winter the waves are stronger and break on the beach face rapidly. Although these waves are strong, they lose a large amount of velocity after breaking. Thus the velocity of the returning water is lower and capable of transporting only the finer (summer) sediments offshore to the bar. In winter, therefore, only coarse sand would be found on the berm, since the winter waves tend to transport the finer sediments offshore to the bar (see Fig. 6-15A and B for typical winter and summer berm profiles).

The horizontal movement of sediment along a coastline is responsible for a myriad of occurrences, ranging from the shoaling of inlets and harbors to the geological aging of a coastline. Indented shorelines can be classified according to geological age: *Type I shorelines,* which are considered to be in their young stage; *Type II shore-*

lines, which are in their middle age; and *Type III*, old shorelines. Barrier beaches are termed Type IV shorelines and, although the basic mechanism for barrier-beach formation is similar, the end result is different.

A Type I shoreline is characterized by rocky headlands extending into the ocean (Fig. 6-17) with a narrow thin beach consisting of only a thin veneer of sand between these headlands. The sediment composing this *pocket beach* has been created in place by the undermining of cliff faces and the grinding of these rocks by wave action. Since the headlands project into the sea, the major wave shock is concentrated on them. This causes the materials composing the headlands to be reduced, initially to rocky fragments and ultimately to sand. Since the waves attacking the shoreline are formed by wind action, these waves will tend to travel onshore from the same general direction as the wind. As the wind continues forcing water shoreward, large quantities of water will tend to accumulate along the cliff face. This water cannot return directly offshore since it would encounter additional incoming waves and be forced back along the cliff face. Consequently, large quantities of water tend to accumulate in the surf zone and ultimately move down the coast in the same general direction as the prevailing wind (see Fig. 6-17). As this water begins to move down the coast it forms into a well-defined current termed a *longshore current*. As the current gains speed it picks up the newly formed

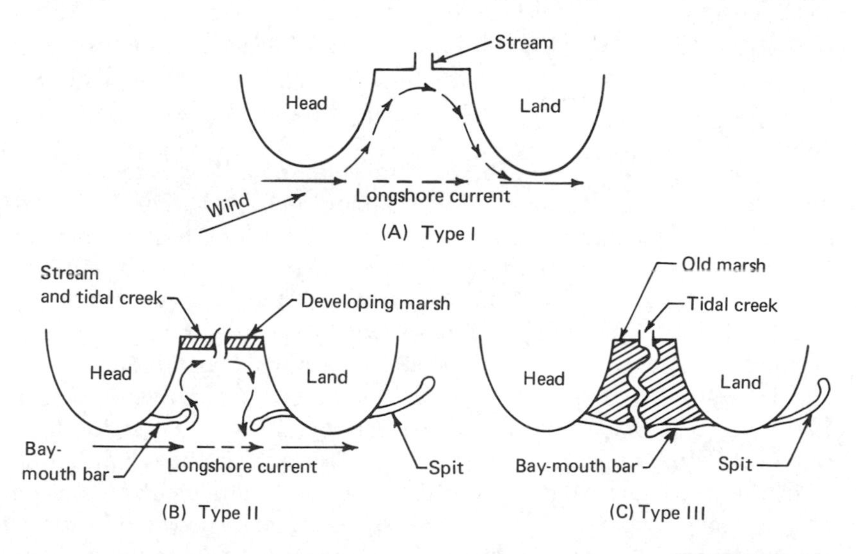

FIGURE 6–17

The erosion of headlands and the formation of bay-mouth bars.

materials that were eroded and reduced to sand from the cliff face and carries this material alongshore as suspended particulate matter. Eventually the longshore current reaches the end of the headland and diverts into the calmer waters to the rear (Fig. 6-17A and B). As it diverts into these areas it loses velocity and the coarser sediments begin to settle out. The current continues to the rear portions, slowing as it travels and dropping finer and finer sediments in the process. Eventually, in the vicinity of the pocket beach, the current is traveling very slowly and, thus, carrying only the finest sediment (silts, for example). The water at this point strikes the pocket beach, slows even further, and releases the remaining suspended sediments. Through this process the pocket beach begins to be covered with the finer materials, and the deeper waters begin to fill with slightly larger sediments. At the point of the original diversion of the longshore current, sediment continues to be deposited. Eventually the coarse sand deposited at the point of initial diversion builds up to such an extent that it breaks the water at low tide and, if left undisturbed, will become permanently exposed and form a bay-mouth bar. The result is twofold: the formation of bay mouth bars extending out into the harbor from the eroding headlands, and a widening and thickening of the pocket beach at the rear of the harbor.

After encountering the pocket beach the water tends to travel along the rear portion of the harbor in the same direction as the longshore current. The velocity of this current will increase since additional incoming water will continue to enter the harbor, deposit new sediment in the same manner as discussed above, and will, consequently, tend to push the water along, eventually forming a well-defined current (Fig. 6-17B). As this current forms, it will move along the opposite shoreline, will increase in velocity, and will tend to remove sediment and carry it toward the mouth of the harbor. This current will encounter the remnants of the original longshore current at the opposite headland. As these two water masses meet, each will lose velocity and sediment will be deposited at the base of the other headland (Fig. 6-17B). This process will ultimately form a smaller "secondary" bay-mouth bar opposite the other headland (with its major bay-mouth bar). Since the current returning seaward is of a lesser velocity than the original longshore current, it carries finer sediments, which are deposited on the "secondary" bay-mouth bar. Thus the "secondary" bar is generally of a finer sediment size than the primary bay-mouth bar. This process is, therefore, responsible for three major occurrences: the building and extension of the bay-mouth bars at the front of the harbor, the tendency of the harbor to become shallower, and the conversion of the pocket beach from a sandy to a silty substrate. Eventually, as discussed previously, the rear portions of this area will build to such a height that they will only be flooded during periods of high tide. The *Spartina alterniflora* will then begin to grow, which will lead to an increased sedimentation rate, an increased raising of these areas, and the replacement of the S. *alterniflora* with the S. *patens–D. spicata–J. gerardii* assemblages.

Once a coastline exhibits permanent bay-mouth bars at the harbor mouths, the area is considered to be in its Type II stage or middle age (Fig. 6-18). During this stage the longshore current continues to build the primary bar, and the returning current formed within the harbor continues to build the secondary bar. As these bars continue to lengthen and heighten, they afford greater and greater protection to the waters behind them. This results in calm conditions and an increased sedimentation rate of finer and finer materials. The rear portions of the marsh tend to build at an accelerated rate and the S. *patens* replaces the S. *alterniflora* in these areas, forcing the S. *alterniflora* to the fringes of the marsh.

As time progresses the salt marsh continues to encroach on

FIGURE 6–18

Extensive bay-mouth bar formations. (*Courtesy Suffolk County Planning Department*)

the open water and the bay-mouth bars continue to build across the entrance. Ultimately, the harbor mouth is effectively sealed off, and an extensive salt marsh covers the entire area behind the bay-mouth bars (Fig. 6-19). The only open water at this stage would tend to be a tidal creek meandering through the extensive marsh lands. When an area reaches this stage it is considered to be in its old age; it is a Type III beach.

Straight shorelines have the same horizontal and vertical movements of sand: However, in these situations, the sand can be transported for considerable distances prior to deposition. Eventually, the longshore current will reach land's end and the water will deepen, causing the current to disperse and decrease in velocity. At this point the sediment will be deposited, forming *spits*. A spit forms in the same manner and by the same mechanisms as bay-mouth bars. However, in these cases, the spit is not "growing" toward another landmass but is, rather, extending out as a peninsula.

FIGURE 6–19

An ancient harbor. Note the extensive salt marsh formation and the meandering tidal creek to the left of the jetty. (*Courtesy Suffolk County Planning Department*)

Barrier beaches (Type IV beaches) are generally formed by the breaching of a spit. This commonly occurs during storms, when a spit is cut off from the mainland by wave action. This will form an inlet between the spit and the mainland that is kept open by current and wave action or by artificial means such as dredging. Regardless of the method of inlet maintenance, the landmass cut off from the mainland is termed a *barrier beach* (Fig. 6-20). Since wind action also forms a longshore current in these areas, there would be a horizontal, downcoast flow of sand here also. This movement of sediment downcoast during storms is responsible for massive beach-stabilization projects, as discussed in Chapter 8.

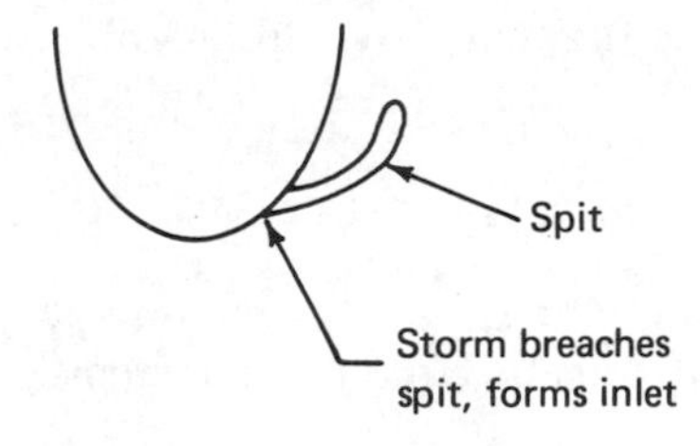

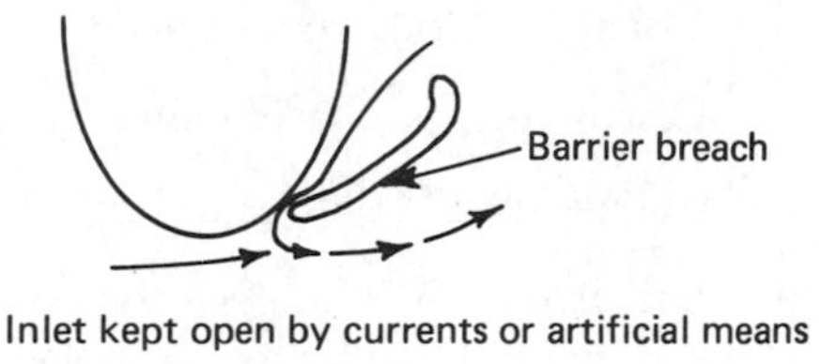

FIGURE 6–20

Barrier beach formation.

SUGGESTED READINGS

Anikouchine, William A., and Richard W. Sternberg. 1973. *The World Ocean: An Introduction to Oceanography*. Englewood Cliffs, N.J.: Prentice-Hall, Inc.

Bascom, Willard. 1964. *Waves and Beaches*. Garden City, N.Y.: Doubleday & Co., Inc.

Benton, Allen H., and William E. Werner. 1974. *Field Biology and Ecology*. New York: McGraw-Hill Book Company.

Groen, P. 1967. *The Waters of the Sea*. New York: Van Nostrand Reinhold Company.

Gross, M. Grant. 1971. *Oceanography*. Columbus, Ohio: Charles E. Merrill Publishing Company.

Smith, Robert L. 1966. *Ecology and Field Biology*. New York: Harper & Row, Publishers.

QUESTIONS

1 / Explain the origin of the dissolved materials in seawater.

2 / Define salinity.

3 / What is the relationship of the law of constancy of composition to chloride-ion concentration and to salinity?

4 / What is the relationship of salinity to density?

5 / Distinguish between pressure gradient currents I and II.

6 / List the various zones encountered in the marine environment.

7 / Discuss the two basic methods of estuarine classification.

8 / Compare and contrast estuarine and lake aging.

9 / Discuss the mechanisms involved in setting water in motion in the lesser, the greater, and the Gulf Stream circulation systems.

10 / Discuss the aging of a given shoreline from a Type I to a Type III system.

CHAPTER 7

types and sources of contamination

As early as 1960 a Senate Subcommittee on Water Resources identified and divided water pollutants into eight standard categories: oxygen-demanding wastes, infectious agents, plant nutrients, organic chemicals, inorganic elements and compounds, sediments, radioactive materials, and heat.

Oxygen-demanding wastes are primarily organic materials that are oxidized by bacteria to carbon dioxide and water. These substances are deleterious because their decomposition leads to oxygen depletion in both freshwater and marine systems. They also produce unaesthetic odors, endanger water supplies, and decrease the recreational value of waterways.

In any community it must be assumed that a certain number of individuals will be diseased and thus capable of contaminating the water with various *infectious agents*. In addition, hospitals dispose of waste products into waterways and groundwater systems. Unfortunately, the identification of specific infectious agents in a water supply requires the analysis of vast numbers of samples by time-consuming and sophisticated methods. Generally, therefore, routine monitoring of water supplies employs the *MPN* (*Most Probable Number*) *Method.* This analysis determines the most probable number of intestinal bacteria that occur in a given water sample. Although these organisms are not pathogenic, their concentration is a fairly reliable indicator of possible pathogenic contamination of a given water supply.

Plant nutrients (phosphorus and nitrogen) enter fresh and marine systems and lead to or intensify eutrophication of these systems (see Chapter 8 for details). These nutrients tend to accumulate in groundwater since it is out of the euphotic zone. As the groundwater moves laterally and reaches the surface waters, these materials add to the nutrient levels already present.

Organic chemicals can be considered to be any compound that contains one or more carbon atoms in its molecular structure. Organics that commonly enter waterways are pesticides, detergents, and hydrocarbons. The term *pesticide* may be applied to any material used to kill pests and covers insecticides, herbicides, rodenticides, and fungicides.

Laundry *detergents,* a common constituent of wastewater, are another organic contaminant. Detergents consist of two major components: a surfactant or sudsing agent and a series of builders. The surfactant lowers the surface tension of the wash water, concentrates, and is preferentially absorbed at the surface and thereby replaces the dirt. After the dirt is removed (replaced by the surfactant) the surfactant is removed in the rinse water. The builders isolate the common elements found in hard water (calcium and magnesium), which would tend to interfere with the action of the surfactant. The surfactant or sudsing agent can become a problem since it is not readily broken down by bacterial action and has a long residence time. Even in relatively low concentrations, it can cause water to form foam (see Appendix III for analytical methods).

Most detergents manufactured in the early 1970s were composed of approximately 40% sodium tripolyphosphate, the "builder" component. It is calculated that such phosphorus-bearing builders account for one half the total phosphate in most wastewaters. Because of this extremely high phosphorus concentration, these detergents are a major source of eutrophication in freshwater systems. In addition, since detergents contribute a large amount of phosphorus to wastewater without a corresponding addition of nitrogen, a nitrogen/phorphorus imbalance will occur when this material enters either marine or freshwater systems.

Hydrocarbons, in the form of gasoline and motor oil, although insoluble in water, are carried from roadways and parking areas in rainwater runoff. Sumps (surface-water catchment basins) accept the water from these areas via a system of storm drains. From these basins the water and associated hydrocarbons percolate down into the water table. Some storm drains empty directly into the nearest surface-water body, carrying the storm water and associated hydrocarbons directly into the waterway.

In addition to the elements mentioned above, analysis of water indicates the presence of a wide variety of other organic chemicals. Some of these compounds are known to be toxic at very low concentrations, while many are not biodegradable or break down very slowly.

Various types of *inorganic chemicals* enter waterways from municipal and industrial wastewater and from urban runoff. These materials decrease water quality, often making it unsuitable for either drinking or further industrial use. In addition, many of these inorganics are not only toxic but tend to concentrate in food chains.

Sediments are washed from the land by storms, floodwaters, and during irrigation. They originate from croplands, overcut forest soils, overgrazed pastures, road-building projects, strip-mining operations, and the like. Estimates indicate that approximately 4 billion metric tons of sediment usually is eroded from terrestrial environments and enter American waterways.

The two other contaminants identified by the Senate Subcommittee were *radioactive substances* and *heat*. Many radioactive substances are lethal at relatively low concentrations and in minute amounts may be mutagenic. Increasing the water temperature of a system is harmful since it generally alters the chemical, physical and biological characteristics of that system. In addition to the possibility of decreasing or eliminating various aquatic forms, it may also stimulate spawning at a time of year when food supplies are limited. This leads to starvation of the newly spawned individuals in the population (see Chapter 8). High temperatures also decrease the density and viscosity of water, causing an increased settling rate of suspended solids. Evaporation rate is increased and, in marine systems, this may lead to localized areas of abnormally high salinity.

sources of contamination

The major sources of water contamination are domestic, industrial and agricultural waste, as well as solid waste, thermal pollution, oil spills, and radioactive waste.

Domestic waste is generated by many sources occupying a large geographical area and consists primarily of sewage from homes and commercial establishments (stores, restaurants, etc.). Generally, the impurities in domestic wastes are diluted and seldom total more than 0.1% of the total mass. This material is largely organic and is oxidized by bacterial decomposition to nitrate, phosphate, carbon dioxide, and water. Since this type of decomposition requires the use of dissolved oxygen, it places an oxygen demand on the system. Because of this tendency to remove oxygen in the decomposition process, a common indicator used to monitor this type of input in receiving waters is the *BOD (Biological Oxygen Demand) Test*. In this analysis (see Appendix III for methods) the amount of oxygen required for decomposition is measured over a 5-day period. In systems receiving significant amounts of organic material, the bacterial decomposition will remove large amounts of dissolved oxygen, which leads to oxygen depletion. These systems are said to have a high BOD. Conversely, when the input of contaminants is absent or minimal, the oxygen levels will not be drastically reduced, since the bacteria will not require an excessive amount of oxygen to decompose small amounts of materials. In these cases there would be a low BOD on the system.

Although approximately 70% of the population of the United States is domiciled in buildings connected to sewage-treatment systems, the majority of these systems are inadequate or inefficient and consist mainly of either primary or secondary treatment (see Chapter 9). These systems bring sewage to a central treatment facility and, after minimal treatment, release this material into waterways, lakes, oceans, or estuaries. This practice of concentrating sewage, subjecting it to minimal treatment, and then releasing the effluent leads to high phosphorus and nitrogen levels, as well as reduced oxygen concentrations in many of the receiving waters. The remaining 30% of the population resides in areas that are unsewered and either releases raw sewage into waterways or relies on cesspools or septic tanks for waste disposal. Sewage disposal by means of cesspools or septic tanks leads to serious localized problems (Chapters 5 and 8). Release of raw sewage into waterways intensifies the problems already discussed in connection with release subsequent to primary or secondary treatment.

Since industrial pollution occurs in large amounts in specific geographic areas, the collection and treatment of these materials should be easy to accomplish. Unfortunately, only a small percentage of the approximately 300,000 factories using and contaminating water in their manufacturing processes adequately treat the water prior to its release. Since the specific type or types of industrial contamination vary with the geography and natural resources of a particular locality, no attempt is made here to correlate water contaminants on a regional basis but merely to identify the possible sources of contamination.

Wastes from textile manufacturing processes are generated from the washing out of the impurities in the fibers, as well as in the discarding of chemicals used in the processing of the fibers. Generally these wastes are organic, have a high BOD, and are extremely alkaline.

Food-processing wastes from meat, dairy, and sugar-beet processing, as well as brewing, distilling, and canning operations, generate large amounts of organic by-products that are disposed of in wastewater. When the wastewater is discarded, along with these by-products, it leads to high BODs and a consequent oxygen depletion in the receiving water via the same bacterial processes involved in the decomposition of domestic wastes.

The effluent released from pulp and paper processing operations is a mixture of chemicals used in the digestion of raw wood chips, cellulose fibers, and dissolved lignin. This wastewater also contains paper and wood preservatives, such as pentachlorophenol and sodium pentachlorophenate, as well as methyl mercaptan, all of which are toxic to fish. This effluent is brownish in color and lowers the photosynthetic rate of aquatic communities by hindering sunlight penetration into the water column (see also Chapter 4). Consequently, the organic wastes from these plants increase the BOD of the receiving water, while color imparted to the water interferes with sunlight penetration, reduces photosynthesis, and further lowers oxygen levels.

Metal industries place a wide array of contaminants in their wastewater. The specific contaminants and concentrations depend solely on the particular manufacturing process employed. For example, steel mills use and contaminate water in the coking of coal, the pickling of steel, and the washing of flue gases from blast furnaces. These waters, after use, tend to be acidic and contain various deleterious substances, such as phenol, cyanogen, ore, coke, and fine suspended solids. Other industries release traces of the metals produced or plated in their wastewater. Metals commonly found in these wastewaters are chromium, mercury, nickel, lead, copper, and cadmium.

A variety of contaminants enter marine and freshwater environments in the effluents released from the various chemical manufacturing plants. The release of acids results not only from acid manufacturing processes but from practically all other chemical manufacturing processes as well. In addition, synthetic fibers (e.g., rayons), bases, pesticides, and other organic and inorganic chemicals are added, depending on the products being manufactured.

Agricultural waste includes the pesticides that are sprayed on crops, as well as sediment, fertilizers, and plant and animal debris that are carried into waterways during periods of rainfall or as runoff and during the irrigation of farmland. Wastes generated by farm animals are also included in this category (Fig. 7-1). Until the mid-1950s animal wastes posed little problem, because they were, for the most part, reused as fertilizers. With the advent of agribusiness, however, the trend has been to ship the animals to large feedlots for fattening prior to marketing. This practice of keeping large numbers of animals in a small area has led to an excess of animal wastes generated in and confined to a given area, where it is economically impossible to distribute wastes for reuse as fertilizers. These materials become a

FIGURE 7–1

Duck farms located along streams and estuaries are a significant source of contamination. (*Photo: G. Marquardt*)

problem when they are allowed to enter waterways during the cleaning of the confinement areas or during periods of heavy rainfall, when runoff carries them into adjacent waterways. Since these wastes are organic, they increase the BOD of the receiving waters.

Inorganic fertilizers, being plant nutrients, lead to overfertilization of waterways when they enter these systems through runoff or during irrigation. The addition of excess plant nutrients can lead to a disturbance of the phosphorus/nitrogen balance in these systems (Chapter 4), as well as excessive plant growth. When the plants die, they settle to the bottom and, since they are organic, increase the BOD of the system during decomposition.

Soil erosion poses a fourfold problem. It increases the normal rate of filling of the waterways into which it washes, decreases the amount of fertile land for crop production, carries pesticide-coated soil particles into the water, and decreases the transparency of the water, which limits photosynthesis. In addition, the sediment carried into freshwater systems tends to clog the gills of adult fish and settles out over incubating eggs, causing suffocation.

A remarkably large number of pesticides have come into wide-

FIGURE 7–2

Agricultural operations place large amounts of nutrients and pesticides into waterways. (*Photo: G. Marquardt*)

spread use in recent years. Many of these compounds are not only nonbiodegradable but are also only slightly soluble in water. Consequently, when sprayed on cropland they remain in the soil for long periods of time. During periods of heavy rainfall or when the crops are irrigated, they tend to be carried, as suspended particles, into surface, marine, or groundwater systems (Fig. 7-2). In both fresh and marine systems they enter the food chain, undergo concentration in nontarget organisms, and increase in animal tissue to alarming levels. In surface, fresh, and groundwater systems they may also enter the drinking-water supplies of various communities.

Solid waste varies in composition with the socioeconomic status of the generating community. The following materials may be classified as solid waste:

1 / Garbage, which includes all decomposable wastes from households, as well as from food, canning, freezing, and meat-processing operations that are not disposed of in wastewater.

2 / Rubbish includes all nondecomposable wastes. These materials may be either combustible or noncombustible. Combustible

materials would include garden wastes, cloth, and paper. Noncombustible materials include masonry, some chemicals, metals, and glass.

3 / Sewage sludge is generated from the settling processes in primary, secondary, and tertiary treatment methods (see Chapter 9), as well as the solids from cesspools, which must be removed periodically.

4 / Miscellaneous materials include industrial wastes, such as chemicals, paint, and explosives, as well as mining wastes, such as slag heaps and mine tailings.

The disposal of solid waste poses many problems, depending upon both the type of waste and the disposal method employed. The majority of the waste classified as combustible–rubbish, garbage, and sewage sludge–is disposed of by one of three major methods: incineration, using it as landfill, or disposal by ocean dumping. Incineration generally leads to air pollution; landfill operations or ocean dumping lead to water contamination. If landfill disposal is used, the material, as it decomposes, will dissolve in or become suspended in the rainwater percolating into the ground and thus into subsurface aquifers. This tends to contaminate not only groundwater but also surface waters, since the two systems are ultimately interconnected (see Chapters 4 and 5). New York City, as well as numerous other coastal communities, commonly disposes of its solid waste by ocean dumping. Because of the highly organic nature of this material, a large BOD is placed on the receiving waters, and the sediment becomes coated with a highly organic ooze. Noncombustible materials are generally disposed of in landfill sites or by ocean dumping.

Thermal pollution occurs because many electric-generating companies use water in the process of cooling their generators. This heated water is then released into the system from which it was drawn, causing a warming trend of the surface waters. Thermal pollution results when the heated effluent is released into poorly flushed systems. In these cases permanent temperature increases often result, which tend to decrease the solubility of dissolved oxygen. In lakes it is also possible to bring about nutrient redistributions (Chapter 8) and prolong summer stagnation periods (Chapter 4).

When heated water is released into large, well-flushed marine systems there is little if any permanent temperature rise. There are, however, problems related to the operation of plants utilizing marine waters in the cooling process. Evidence indicates that seawater tends to corrode the cooling pipes, which are generally constructed of a copper–nickel alloy termed Monel. These metals readily dissolve in the heated seawater and are then released into the marine environment together with the heated effluent. This adds to the nickel and copper concentrations of these systems. In addition, the screens covering the water-intake pipes rapidly foul with marine organisms, which decreases the flow of water into the plant. The screens are commonly cleaned by using a concentrated detergent solution or copper sulfate. These cleaning materials are then released into and contaminate the surrounding waters.

Oil pollution results from accidents involving oil tankers and from spills at offshore oil drilling sites. A more persistent source of oil pollution results from the practice of oil tankers, after they deliver the oil, to fill the empty tanks with seawater to act as ballast for the return trip. Prior to docking, the seawater ballast, contaminated with the oil that remains in the tanks, is discharged. Although this practice is illegal, it is difficult to prevent. The alternative is to pump the contaminated seawater into tanks at port. This is not only inconvenient but uneconomical. In addition, there will remain the problem of disposal when the disposal tanks are full.

The major sources of radioactive wastes are nuclear explosives, accidents at nuclear power plants, fuel-reprocessing plants, and research laboratories and hospitals that release these wastes into the atmosphere or into wastewater. Presently, most interest centers on radioactive iodine and strontium, since man is at the end of the food chains that concentrate these elements. Much more research is necessary before the implications of long-term exposure to low-level radiation can be determined.

The significance of the various parameters discussed and their use as water-quality indicators is summarized in Table 7-1.

TABLE 7-1

water-quality indicators

Parameter	*Significance*	*Level*
Dissolved oxygen	General indicator of water quality; source of O_2 for respiration	Minimum acceptable level, 4–5 mg/liter; 10–15 mg/liter for reproduction of desirable fish
Total suspended solids	Clog fish gills, bury eggs, reduce light penetration, increase heat absorption	Dependent on location
Total dissolved solids	Represents total mineral content which may or may not be toxic	A maximum of 400 mg/liter for diverse fish populations

TABLE 7-1 (cont.)

Parameter	*Significance*	*Level*
BOD	Amount of dissolved oxygen removed during decomposition of organic matter in a given time; a general indicator of contamination due to biodegradable organics	*BOD* *Water Status* 1 mg/liter Very clean 2 mg/liter Clean 3 mg/liter Fairly clean 5 mg/liter Doubtful 10 mg/liter Contaminated
COD	Indicates the concentration of materials oxidizable by chemical reaction	0–5 mg/liter indicates very clean streams
pH	Indicates the addition of acids or bases	pH depends on actual system
Iron	Excessive amounts can clog fish gills; indicates drainage from iron-bearing sediments, mines, industrial processes	A maximum of 0.7 mg/liter for diverse fish populations
Manganese	Concentration low in natural systems due to low solubility; high concentrations indicates contamination	A maximum of 1 mg/liter is a common criterion for stream quality
Copper	Indicates drainage from copper-bearing sediment, mines, plating, or other industrial sources	A maximum of 0.02–10 mg/liter is a common criterion for stream quality
Zinc	Indicates mine drainage or industrial input	A maximum of 1 mg/liter is a common criterion for stream quality
Hg, Cd, Pb, Ni, Cr, Co, Ag, etc.	Indicates industrial input	A maximum of 1 mg/liter is a common criterion for stream quality
Nitrate	A major plant nutrient; in high concentrations it can promote excessive plant growth; major sources are fertilizers, sludge, and sewage	A maximum of 0.3 mg/liter to prevent excessive fertilization of streams
Phosphate	A major plant nutrient; major sources are detergents, fertilizer, sewage	A maximum of 0.03–0.40 mg/liter total inorganic phosphate is a common criterion

SUGGESTED READINGS

Ehrlich, Paul R., and Anne H. Ehrlich. 1972. *Population, Resources, Environment.* San Francisco: W. H. Freeman and Company.

Hodges, Laurent. 1973. *Environmental Pollution.* New York: Holt, Rinehart and Winston, Inc.

Warren, Charles E. 1971. *Biology and Water Pollution Control.* Philadelphia: W. B. Saunders Company.

QUESTIONS

1 / List the eight standard categories of water pollutants.

2 / Explain why plant nutrients may be considered a pollutant.

3 / Explain how heat may cause deleterious effects.

4 / What are the major sources of water contamination, and why?

5 / Explain the term BOD.

CHAPTER 8

consequences of pollution

A pollutant is considered to be an undesirable or deleterious modification of the environment. The modification may actually or potentially affect human life, living conditions, cultural assets, or the life cycles of the indigenous plant and/or animal communities that inhabit a given system. The major sources of pollution, discussed in Chapter 7, are considered to be domestic, industrial, agricultural, radioactive, and solid wastes, as well as thermal pollution and oil spills. These contaminants may be conveniently subdivided into four categories on the basis of their effects on a given system regardless of their source: (1) substances that lead to oxygen depletion, (2) excess plant nutrients, (3) agents of biological dysfunction, and (4) sedimentary and erosional processes.

A given contaminant may fall into one or more of these categories depending on many factors, such as its mode of action, the amount of dilution it encounters as it travels from the point of input, the flushing rate of the system into which the contaminant is released, and the tolerances of the organisms encountered. Each contaminant must, therefore, be evaluated in terms of the particular characteristics of the receiving water. For example, the consequences of releasing a small amount of raw sewage into a shallow, artificial pond would be totally different from the release of this material into the deep ocean. The differences are one of degree and are due not to the inherent differences between marine and fresh water but rather to the differences in volume, circulation patterns, and degree of dilution. Thus a shallow, poorly flushed pond could be expected to behave in a fashion similar to a shallow, poorly flushed estuary when a contaminant is introduced.

Not only must each contaminant be evaluated in terms of the particular characteristics of the receiving water, but the effects of synergism must also be considered. Synergism is defined as a combination of factors (in this case, contaminants) that reinforce the activities or effects of each other. Synergistic effects increase the impact of a contaminant, since the total effect from the interaction of these contaminants with other materials that may be present within the environment, or within a specific organism, is often greater than the effects of each individual contaminant. Thus it is often impossible to evaluate the effects of individual contaminants on a given system since, in many cases, the combined effects of two or more contaminants are more severe than identical concentrations of a single contaminant. For example, cyanides in water, while toxic to aquatic life, are extremely lethal in the presence of cadmium and/or zinc. Thus the presence of cadmium and zinc is said to have a synergistic effect on the toxicity of cyanides.

Since marine and surface freshwater systems are generally affected similarly by the addition of deleterious materials, this discussion will encompass the effects on both of these systems. Groundwater systems are subjected to completely different ecological factors (total absence of a euphotic zone, for example), are affected in a different manner by the input of various contaminants, and will be considered separately. In addition, it is to be noted that this section deals only with effects on natural communities and systems. It does not, for the most part, discuss the implications of and effects on man. Human effects are discussed on p. 128.

reduction in oxygen levels

Factors as diverse as oil spills, heat, suspended sediment, organic wastes, and some inorganic wastes are known to decrease the available oxygen in a given system. Although the end result, oxygen depletion, is the same, the mechanisms leading to oxygen depletion vary. There are four major methods that may serve to reduce the oxygen levels within a given system: (1) decreasing the photosynthetic rate of the plants, (2) decreasing the solubility of the oxygen within the water column, (3) interfering with the diffusion of atmospheric oxygen at the air–water interface, and (4) increasing the oxygen consumption of the aerobic bacterial component of the system (increasing the BOD).

decreased photosynthetic rate

Photosynthetic rate can be altered by increasing the turbidity of the water column, thus decreasing the amount of light that enters a given water column. This may occur from suspended particles that enter a given system either by erosion during periods of rainfall or during the irrigation of crops. This material enters the water, physically blocks the amount of light entering the system, and thereby decreases the amount of light available for photosynthesis. Since all plants require sunlight to provide the energy necessary to carry on photosynthesis, excessive addition of sediment will reduce oxygen levels within the water column and, if prolonged, may eliminate the plant life altogether. This would have the added effect of eliminating the food source of the primary consumers. The subsequent death and decomposition of these organisms would increase the decomposition rate in the regeneration zone and further increase the BOD on the system. Light may also be physically blocked from the water column by the input of highly colored effluent from dyeing processes in the textile industry and from paper-mill effluents. The textile industry releases a variety of highly colored effluents, while paper-mill effluents tend to be a deep brown in color. Many of these highly colored effluents decrease light penetration through the water column, and thereby interfere with aquatic productivity in both freshwater and marine environments.

Photosynthetic rates may also be reduced by directly eliminating the plants that produce the oxygen. Mercury, which enters the environment from a variety of sources (discarded electrical batteries, mercury-based pesticides, mining operations, etc.), travels to the regeneration zone, where, by bacterial action, it is converted to the organic methyl mercury, which is soluble in water. After it is in its soluble form, it is capable of entering the food chain through uptake by phytoplankton or by direct assimilation by filter feeders. Mercury is highly toxic to phytoplankton. It has been found that photosynthesis is inhibited at mercury concentrations of 0.1 parts per billion (ppb), which is one fiftieth of the permissible U.S. Public Health limit of 5 ppb. Many pesticides (in addition to those containing mercury) are also known to kill phytoplankton, thereby reducing photosynthetic rates, and thus oxygen levels, in the systems they enter.

decreased oxygen solubility

Oxygen levels are also decreased as the temperature of a given system is increased. Increasing temperatures tends to increase the molecular motion of the water and any dissolved gases, which decrease the solubility of the dissolved oxygen. Lakes are particularly sensitive to increased water temperatures, since this reinforces the temperature–density barrier (Chapter 4) and prevents efficient mixing of the surface euphotic zone and the hypolimnion. Thus, if the temperature input is great enough to prevent or delay the normal fall overturn, severe and prolonged anaerobic conditions will occur in the hypolimnion. This will result in overkills of hypolimnetic populations. In addition, higher temperatures are favorable to increased bacterial growth and also increase the metabolic processes of the bacteria. This has the net effect of increasing the decomposition rate in the regeneration zone. Since bacterial decomposition (aerobic) requires oxygen, an increase in the rate will also tend to increase the depletion of oxygen levels.

decreased diffusion of oxygen

Substances that interfere with the diffusion of oxygen at the air–water interface by blanketing the water surface and physically preventing oxygen from entering the system also reduce the amount of available oxygen. Oil entering the environment from a variety of sources decreases oxygen levels in this manner. For example, the *Torrey Canyon* disaster of 1967 released 117,000 tons of crude oil. This oil blanketed large areas of the marine waters and, in addition to reducing oxygen diffusion, destroyed entire seabird colonies, harmed other marine organisms, and eventually coated both British and French beaches, temporarily ending their use as recreational facilities. A less-well-known source of oil pollution is the disposal of used motor oil. It is estimated that 350 million gallons of used motor oil are generated per year. Since it is economically unfeasible to re-refine this oil, it is treated as a waste product and is eventually poured into municipal sewage systems, into storm drains, or directly into a convenient waterway. Ultimately the material released into sewage systems or into storm drains finds its way into streams, lakes, or estuarine waters, where it behaves similarly but on a reduced scale to the oil spilled by the *Torrey Canyon.* It is to be noted, however, that the used motor oil is not released into large marine systems, such as the English Channel, but rather into small systems. Thus it is possible that this material, discarded almost continually into small, poorly flushed waterways, has a greater total impact on these systems than the more spectacular large spills.

The leaves of deciduous hardwoods, falling into streams in the autumn, can reduce oxygen levels by hindering sunlight penetration as they float on the water's surface, and can interfere with the exchange of oxygen at the air–water interface. In addition, the leaves eventually sink, reach the regeneration zone, and decompose, thereby increasing the BOD on the system. These effects are important when they occur during periods of reduced stream flow (generally in the autumn of a dry summer). At these times a combination of reduced oxygen levels and low water volume will be responsible for lowered oxygen concentrations in localized areas. These effects are transitory, however, and seldom so long prolonged as to completely reduce oxygen levels over a long period of time. In addition, all these materials are biodegradable and do not cause prolonged, unaesthetic results.

increased oxygen demand

The oxygen demand (BOD) of a system can be increased by the addition of both organic and some inorganic substance to the environment. Some mechanisms have already been discussed as side effects to decreasing oxygen levels by other mechanisms. In general, organic contaminants entering systems from municipal sewage-treatment plants or as raw sewage (New York City presently releases over 1 billion liters of raw sewage into the Hudson River, which is estuarine at the various points of input), as well as animal wastes from feedlot cleaning operations and plant and animal residues from food processing operations, are the major source of organic wastes entering our waterways. All of this organic material eventually reaches the regeneration zone, where it can be broken into its component parts through bacterial action. Initially this is accomplished by aerobic bacteria, which require oxygen to perform the decomposition process (see Chapter 2). Since there is a large amount of organic matter to be broken down, the bacteria remove large quantities of dissolved oxygen from the system (the BOD increases). As additional material is added to the regeneration zone, more oxygen is removed, thus decreasing the oxygen levels further and accelerating the problem. If the input of material continues for protracted periods (as it does in the vicinity of industrial or municipal outfalls) the oxygen levels will decrease drastically and the system will become permanently anaerobic in these areas.

In most natural systems anaerobic conditions reach a peak a short distance downstream from the point of input. This point coincides with the disappearance of the normal plant and animal communities and the appearance of large populations of undesirable or unaesthetic organisms that are able to prosper under these largely anaerobic conditions. Typical organisms found in these areas are species of gas-producing bacteria that reduce sulfate (from paper-processing operations), the increasingly common white sewage fungus (actually a species of bacteria—*Sphaerotilus natans*), the protozoan that preys on bacteria—*Paramecium putrinum,* and the small thread-like sewage worm *Tubifex tubifex.* In areas subjected to such strenuous environmental conditions the species diversity is low, but the populations of the few adaptable species are large. Unfortunately, as mentioned above, these organisms are undesirable and/or unaesthetic and can rarely be utilized as food. Fish populations are either totally eliminated if the anaerobic conditions are widespread, or will tend to migrate to uncontaminated regions. In either event these organisms are absent in the areas of immediate contamination.

Inorganic materials such as iron (ferrous) salts from mining drainage operations and sulfides from pulp and paper processing plants also decrease the oxygen levels in fresh and marine environments. These inorganics decrease oxygen levels by utilizing the dissolved oxygen in processes that serve to convert these materials into different inorganic chemical forms. Generally, the addition of inorganics is said to place a *chemical oxygen demand* (*COD*) on the system.

In most of our rivers today the water is just barely acceptable before it is utilized by a downstream community, being pretreated prior to distribution to this community, and then is recontaminated and treated prior to release back into the river. The rivers seldom have an opportunity to naturally and totally purify themselves before the water is reused, recontaminated, and rereleased farther downstream. Thus the river system is generally low in oxygen, and undesirable and/or unaesthetic organisms are generally found throughout its entire course. Lakes are even more susceptible to contamination, since they are poorly flushed and any materials that are added will tend to accumulate and eventually build up to high concentrations.

excessive plant nutrients

The addition of plant nutrients (phosphorus and nitrogen) to marine and surface freshwater systems will have a number of deleterious effects. In addition, it is to be noted that the organic materials discussed above will decompose into their constituent forms. A portion of their products of decomposition are the plant nutrients: inorganic phosphate and inorganic nitrate. In addition to their effects on oxygen levels, these materials will also add to the problems caused by excessive fertilization discussed in this section. These effects are commonly termed *eutrophication.*

In Chapter 4 it was noted that in surface freshwater systems phosphate was commonly the limiting factor, while in marine systems nitrate tends to limit plant growth and, thus, overall productivity. Consequently, the material that is likely to lead to excessive over-fertilization is the substance that is commonly limiting in undisturbed systems. In other words, the addition of phosphate to many freshwater systems would lead to excessive plant growth, since, in normal freshwater systems, phosphorus is generally limiting. In marine systems the reverse would generally be true. Since nitrogen is limiting, the addition of contaminants containing nitrogen would remove the limiting factor and excessive plant growth would occur.

Overfertilization of a system is harmful since it destroys the integrity of the community by rapidly altering nutrient relationships and plant–animal interrelationships that have slowly evolved over the centuries. The end result is generally the elimination of many or all of the normal populations, which are replaced by a few, opportunistic forms that are more tolerant of the rapidly changing conditions. Regardless of the system, overfertilization tends to remove the limiting factors and allow for the proliferation of and abnormally high growth and reproductive rates of plant populations. If the plants normally found in the disturbed system multiply rapidly, the consumer organisms cannot keep pace with this rapid increase. Therefore, more and more plants will remain uneaten by the consumers. These plants will ultimately die and add excessive material to the regeneration zone to be decomposed. If the source of overfertilization continues unabated, there will tend to be an excessive increase in the materials entering the regeneration zone. This will ultimately place an excessive demand on the dissolved oxygen of the system, and oxygen levels will decline. Many fish are extremely sensitive to reduced oxygen levels, and they will tend to be eliminated. In addition, the accelerated rate of settleable organic matter will increase the rate at which the bottom sediments are covered with highly organic silts and muds. This material will tend to cover and smother many benthic forms adapted to living on sandy substrates, in addition to suffocating eggs and larval forms of many of the normal animals. In some cases excessive fertilization tends to foster the growth of new plant species that did not occur in the area previous to this type of disruption. These new forms often tend to outcompete the indigenous forms in these altered conditions and eventually eliminate them. In many cases the primary consumers are unable to utilize these new forms as a food source and are therefore eliminated from the system, causing grave consequences to the entire food chain.

It is to be recalled that all systems tend to age naturally. This normal aging is accompanied by a gradual increase in the nutrient levels, the slow conversion of the benthic sediments to a mud bottom, and a normal, orderly succession of the plant and animal communities. Overfertilization tends to "mimic" the normal aging process. The problem, however, is that in these cases conditions are altered rapidly. This prevents an orderly succession in which the community structure is able to slowly change and/or evolve in response to the normally slow changes. In community change, associated with contamination, conditions change so rapidly that entire plant and animal communities are eliminated, allowing undesirable opportunistic forms to occupy their place.

Excessive fertilization is the consequence of the release of improperly treated sewage into waterways as well as animal wastes from feedlot cleaning operations and animal and plant residues released as a by-product of food-processing operations (it is to be recalled that all these materials have also been implicated in the reduction of oxygen levels). Perhaps the greatest source of excess plant nutrients in waterways occurs from the use of inorganic fertilizers on agricultural lands as well as on lawns and backyard vegetable gardens (Fig. 8-1). The problem associated with the use of inorganic fertilizers involves their effects on the soil on which they are spread. In natural soils much of the nitrogen is contained in the highly organic humus. Generally, inorganic nitrogen accounts for 2% or less of the total nitrogen content of such soils; the remainder is combined in the large

FIGURE 8–1

Excessive development of lakefront property is a major factor leading to water contamination due to the input of sewage, fertilizers, and pesticides. (*Photo: G. Marquardt*)

organic molecules of the humus (derived from plant residues, animal manure, etc.). A high humic level provides a favorable medium for the chemical reactions and mineral transport necessary for the growth of food crops. The soil bacteria slowly decompose the humus to form the nitrates and other nutrients required by the plants for optimal growth. In addition, the humus increases the ability of the soil to retain water, thereby decreasing the necessity for excessive irrigation and the erosion problems associated with rainfall. It is to be noted that irrigation practices decrease the amount of available water in a given system, while erosion causes loss of soil from agricultural areas. The eroded soil also enters waterways and decreases the amount of available light for photosynthesis, causing siltation problems.

Humus declines in lands heavily fertilized with inorganic fertilizers. This is due to the failure to return crop and animal residues to the fields from which they were removed by harvesting. These materials, if returned to the fields, would aid in maintaining humus levels. Consequently, as humus levels decline, the nitrogen, with no organic humus to "bind" it to the soil, will tend to leach out as the soil is irrigated. This lowers the soil fertility and necessitates additional applications of fertilizer. Attempts to raise soil fertility by additional fertilization (with no addition of humus) will cause a further decline in humus levels, which will cause the soil to retain less nitrogen and lead to further applications of inorganic fertilizers. It is to be noted that the inorganic fertilizers themselves are not nutrient-deficient, since it had been demonstrated that if plant and animal residues (humus) are supplied along with the inorganic fertilizers, the humic levels of the soil will increase (thus improving the soil quality). Consequently, the materials discarded by the food-processing industry (plant residues) and by the practice of maintaining large numbers of animals on feedlots (manure generation) are the materials that would improve soil quality. Presently, these materials are treated as waste, discarded, enter waterways, and lead or contribute to their degradation.

The use of inorganic fertilizer has increased twentyfold in the past 25 years. This has led to a decline in soil quality in much of the nation's farmland. As humus levels decline and the soil's ability to retain nitrate is decreased, additional application of fertilizer is required to achieve the desired growth rate. This fertilizer tends to rapidly leach from the soil and enter waterways. These practices are primarily responsible for the alarming rise in the nitrogen levels in the nation's waterways.

A classic example of the consequences of the excessive use of inorganic fertilizers is the eutrophication of Lake Erie. The waters draining from the farmlands in this region ultimately flow into the lake and have an estimated nitrogen content that has been calculated to be the equivalent of the sewage generated by 20 million people (the total human population of the Lake Erie drainage basin area is 10 million). This nitrogen-rich runoff has altered the entire nitrogen balance of the lake and has encouraged the growth of certain algae that tend to grow rapidly under these conditions, cover large areas, and foul beaches. When they die, the bacterial decomposition of this material places an excessive BOD on the system, which leads to extensive fish kills.

An example of estuarine eutrophication in which the normal nitrogen cycle became "short-circuited" occurred in the Great South Bay off the south shore of Long Island. In the early 1950s concomitant with the postwar building boom, nitrogen levels in the form of urea, uric acid, and ammonia began to increase in the bay. This upset the normal nitrogen–phosphorus ratios (Chapter 6) and led to a complete change in the planktonic communities that inhabit the area. The normal phytoplankton populations consisted of diotoms, flagellates, and dinoflagellates. These populations were almost totally replaced by small flagellates of the genera *Nannochloris* and *Stichococcus*. The new, opportunistic forms were totally undigestible by the bay's primary consumers, and this led to the virtual elimination, by starvation, of the Blue Point oyster in the Great South Bay. Almost all other forms of shellfish were similarly affected, and the majority of attempts to reintroduce them have, to date, failed.

Subsequent laboratory investigations indicated that both *Nannochloris* and *Stichococcus* are able to utilize nitrogen in the form of urea, uric acid, and ammonia (see normal nutrient cycles, Chapter 4), while the normal phytoplankton require nitrogen as nitrate. Thus the opportunistic species "short-circuited" the nitrogen cycle by removing this nutrient prior to its conversion to nitrate. By doing so, the normal phytoplankton were deprived of their nitrogen source and were eliminated. This is an example of an opportunistic species, normally rare in a system, proliferating and taking over when systems become and are allowed to remain in a disturbed state.

Phosphate levels in surface waters have increased 25 fold within the past 15 years. Since they are also a plant nutrient, they have the same effects as nitrogen on the systems that they enter. The major sources of phosphates are fertilizers, agricultural wastes, and municipal sewage. It is calculated that 60% of the phosphate entering U.S. waterways is from municipal sewage. The primary source of phosphate in municipal sewage is from detergents, which in addition to causing overfertilization, cause the familiar foaming now common on many waterways.

The eutrophication problems due to nitrates in Lake Erie are compounded by the addition of phosphorus from municipal wastes. The addition of phosphates, coupled with the addition of nitrate, has caused this system to deteriorate rapidly. Even if the addition of phosphate were to be eliminated, the problem would not be rectified, since a large concentration of phosphate-rich sediments has accumulated over the years. Nutrients tied up in bottom sediments are normally prevented from mixing with the water by a protective layer of insoluble iron salts. When the bottom waters become low in oxygen (a normal occurrence in summer stagnation), the protective layer disperses and phosphate is released into the water.

agents of biological dysfunction

For the purposes of this text, *agents of biological dysfunction* will refer to any contaminant that either directly kills organisms or that interferes with their metabolic or physiological activity or their genetic or reproductive capabilities in such a manner as to threaten the success of a natural population in a given system. In addition, any material that appears to be harmless or to have a negligible effect when ingested, assimilated, and so on by members of one population but tends to accumulate in the tissues of these or of other organisms as it "passes up" the food chain to ultimately affect the success of higher consumers would also fall into this category.

The major materials in this category are the various persistent pesticides, heavy metals (primarily mercury, cadmium, lead, and copper), the by-products of the plastics industry (PVCs and PCBs), heat, and radioactive wastes. By strict definition, many of the materials previously discussed could be considered to fall into this category, since they, too, have deleterious effects on natural populations. However, in this section, only those materials directly related to metabolic, physiological, genetic, or reproductive interferences will be considered.

Perhaps the most alarming and well-documented effect of many of these substances is their tendency to increase to significant concentrations in the various organisms as they pass up the food chain. This phenomenon is termed *biological magnification,* and is known to be responsible for such diverse occurrences as eggshell thinning in the osprey, gross birth defects in terns, and unacceptably high DDT levels in the milk of human mothers.

DDT, a chlorinated hydrocarbon, provides the classic example of biological magnification. Although it is only slightly soluble in water, it is extremely soluble in organic fatty tissues. Algae, in the process of assimilating dissolved materials from the water, indiscriminantly take in and accumulate DDT. The DDT concentrates in their cellular fat bodies and increases on the order of parts per million. Primary consumers (zooplankton, shellfish, etc.), each feeding on tens or hundreds of thousands of algae during their life cycles, will accumulate DDT in the tens of parts per million. The secondary consumers (small fish) that feed on the zooplankton will further concentrate this material and when they, in turn, are preyed upon by the larger fish, these tertiary consumers (since they are feeding on thousands of smaller fish) will concentrate the DDT to significant levels prior to passing it on to the top consumers. Thus this material tends to concentrate in animal tissues at each succeeding step in the food chain, and its effects, if not immediately lethal to the lower consumers, are generally evident in the higher consumers only after many months or years of concentration.

chlorinated hydrocarbons

The other chlorinated hydrocarbons (aldrin, dieldrin, benzene hexachloride, etc.) can be assumed to follow similar environmental pathways. Recently, another class of chlorinated hydrocarbons, the polychlorinated biphenyls (PCBs) have been found to have serious environmental effects. This material is used extensively as heat-exchange and insulating fluids in high-voltage electrical equipment. In addition, it is added as stabilizers to paints, plastics, and rubber to make these materials resistant to decomposition processes. Approximately 4000 tons of PCB enters waterways through sewage, leaching of lubricants, and heat-exchange fluids, as well as from landfill operations. An additional 2000 tons/year is estimated to enter the atmosphere from plasticized materials. PCBs, like DDT, undergo biological magnification when they enter food chains. In the Great Lakes, coho salmon have PCB levels above 5 ppb. The toxicity of this material varies with the species of fish. For example, it is toxic to bluegills and catfish in concentrations above 20 ppm; lethal levels for trout are 8 ppb and for shrimp as low as 1 ppb. In the osprey (top consumers feeding on fish), concentrations of 1000 ppm are found.

Recent studies at a tern nesting colony on Gull Island, New York, at the entrance to the Long Island Sound, indicate that PCBs are responsible for gross birth defects that are appearing with increasing frequency in the tern colony. Workers from the American Museum of Natural History have noted newly hatched chicks with four legs, others with undeveloped legs and feet, or with rudimentary and displaced upper mandibles (bills). Another chick was noted that had no left eye and a crossed bill. This evidence indicates that while the PCBs follows the same pathway of biological magnification exhibited by the other chlorinated hydrocarbons, the end result is different in terns.

metals

Metals commonly find their way into waterways after they are disposed of in various industrial processes. The effects of many of these metals (copper, lead, and mercury) are well known. The effects of other metals (such as chromium, cadmium, cobalt, and nickel), the environmental pathways that they may take, and their effects on aquatic and human populations are less well known. However, the fact that they are found in human food supplies (clams, oysters, etc.), or in the food source of these organisms, is a strong argument for halting further addition of these materials to waterways until their effects have been determined.

Lead and mercury have been two of the most widely studied metals. It has been estimated that in 1974 over 20 million pounds of mercury was released into the environment. The Saskatchewan River in Alberta, Canada, had mercury levels of 0.05 ppb upstream of the city of Edmonton and levels of 0.12 ppb downstream of the city. This example obviously indicates that man is an important factor in increasing the mercury levels of the various waterways. Since these streams ultimately enter into the coastal environment, it is to be expected that this material will eventually enter these systems in ever-increasing amounts.

The pathway of mercury in the environment has been discussed previously (metallic mercury entering the regeneration zone, where it is converted to the soluble methyl mercury). Metallic mercury is relatively nontoxic to organisms; however, once it is in its organic form (methyl mercury), its toxicity increases markedly. Since the bacterial conversion of mercury to methyl mercury is continuous, the metallic mercury added to the various systems will serve as a methyl mercury reservoir that will continue to yield the more toxic form even should the addition of mercury be halted. The effects of methyl mercury on phytoplankton populations have been discussed previously. Methyl mercury is known to be concentrated by other organisms and to undergo biological magnification in the food chain. Fish, for example, tend to both absorb it directly through their gills and also receive additional mercury incorporated in their food. Since methyl mercury is more easily absorbed and more slowly excreted than the inorganic mercury, it magnifies to extremely large concentrations in fish tissue. Pickerel from the Great Lakes have mercury concentrations as high as 5 parts per million. This is approximately 3000 times the concentration of the waters from which they are taken. In addition to decreasing the vitality of these aquatic populations, mercury, when ingested by man, can lead to blindness, deafness, insanity, and/or death.

Lead, a remarkably well distributed element, is even found in the polar ice caps. Analysis of lead concentrations in the annual ice layers deposited on the Greenland and Antarctic ice caps showed an increase in lead concentrations from less than 0.001 microgram per kilogram (μg/kg) deposited at the 800 B.C. level, to 0.03 μg/kg deposited at the 1815 level. These concentrations soared to over 0.20 μg/kg in the present layer. The sharpest increase was determined to have occurred since 1940, which is coincident with the sharp rise in lead additives in gasoline.

Presently, the sources of lead contamination are seemingly endless, and vary from industrial input to the solder used to seal food cans. Evidence indicates that most of the lead enters aquatic systems as particulate lead washed from the atmosphere during rainfall. Numerous studies show that lead is absorbed by phytoplankton or filter feeders and then concentrates as it enters and passes up the food chain.

As far back as 1898 copper was recognized as the metal responsible for imparting a green or blue color to oysters in both England and the United States. This material also enters from a variety of sources and is taken up by the phytoplankton and passed up the food chain. Eventually, copper levels reach sufficient concentrations to impart color, to impair metabolic processes, or to kill various organisms (copper is known to be toxic to lobsters in concentrations as low as 1 μg/liter).

Other metals, such as cadmium, cobalt, and nickel, are also increasing in estuaries such as the Long Island Sound, New York. Since the significance of these levels and their environmental pathways are presently unknown, they will not be discussed here save to note that their presence has also been detected in both clams and oysters; both of these species are utilized as a human food source. Since there have been cases of human illness associated with the excessive intake of these materials and since they are capable of increasing to high levels in human food sources, further additions of these materials to waterways should be halted.

heat

As discussed previously, the solubility of oxygen decreases as water temperatures increase. In addition, the respiration rates of all organisms (including the bacteria) double with each 10°C rise in temperature, which places an added BOD on the affected system. Temperature effects also tend to be synergistic. For example, carp at 1°C are able to tolerate CO_2 levels of 120 ppm, while at 36°C, CO_2 concentrations of 55 to 60 ppm are lethal. In other words, the tolerance of carp to CO_2 is decreased as temperature increases. Similar effects are known to occur with other toxic materials.

Temperature also affects reproductive behavior in many aquatic and marine organisms. Many organisms are induced to spawn prematurely in response to increased water temperatures. Consequently, when water temperatures are raised, the normal spawning sequence is disrupted. In many instances when organisms are induced to spawn at abnormal times (too early in the season, for example) due to the artificial raising of the temperature of a given system, the normal food supply required by the young may not yet be available. This could lead to the elimination of the young individuals in a given area.

radioactive wastes

Radioactive wastes may be directly lethal to organisms or they may lead to genetic disruption. Evidence also indicates that some radioactive materials (e.g., strontium) may be magnified as they pass up the food chain. Elaborate safety precautions have, thus far, prevented significant concentrations of radioactive materials from entering the environment. The projected increases in the use of this material in the future will pose additional problems for the safe transportation, storage, and disposal of this material.

sediments and erosion

Sediments, as discussed previously, have a variety of deleterious effects on waterways; they enter waterways and interfere with photosynthesis by hindering light penetration through the water column. Sediments also clog the gills of fish and settle out and blanket the normal bottom sediments, eliminate spawning areas, and smother the eggs of many species of marine and freshwater organisms. In agricultural areas they tend to be coated with the various pesticides that are applied to crops. When these sediments are carried into waterways by erosional and/or irrigational process, they also carry in the pesticides and add to the pesticide levels of these systems. In addition, sediments ultimately settle out in waterways, behind dams, and so on, and interfere with navigation and/or water storage, necessitating the frequent dredging of these areas. Two other effects involving sediments and related to unwise management practices are coastal erosion control and salt-marsh landfill operations.

coastal erosion

The interference with coastal longshore transportation (Chapter 6), although not a consequence of pollution, is a consequence of unwise management practices and will be considered here. The longshore current (Chapter 6) is formed by wind striking a shoreline on an angle and setting water in motion in the same general direction as the wind. Using Fig. 8-2 as an example, the longshore current will, in this case, travel from east to west carrying suspended sediments. Sediment is picked up at points A, B, C, and so on, and moved downcoast. Sediment from point A replaces sediment removed from point B, and sediment from point B replaces the sediment removed from point C, and so on. It is to be noted that point A is the only area of continuous, net erosion along the entire coastal system.

At any given time a storm may strike the coastline and focus its major force at one particular point. Since the waves will strike the beach at this point with increased force and frequency, the beach will be temporarily cut back at this point. After the storm passes, however, the normal sediment flow will be reestablished and the beach will rebuild.

Problems arise when building is permitted on an unstable area, such as a barrier beach, and are compounded when the structures are erected too close to the high-water line. Prior to the construction

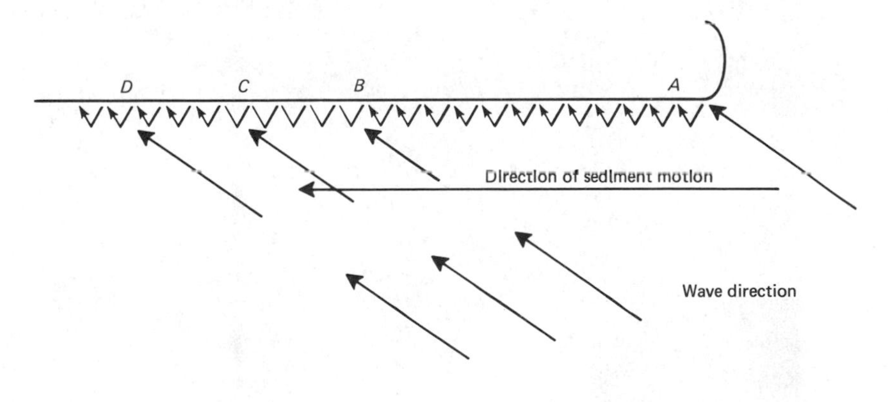

FIGURE 8–2

Sediment picked up at point *A* is moved down the beach in a series of movements from beach to surf to beach by the longshore current.

of residences or commercial establishments, storms removed seemingly insignificant amounts of sediment. Once a premanent structure is present against which sediment loss can be directly measured, the sediment loss becomes alarming (at least to the owner of such a structure). In order to protect private property, demands are soon made for "beach-stabilization" projects. This results in the construction of *groins* to rebuild the beach by slowing the longshore current (Fig. 8-3). Using Fig. 8-4 to illustrate, the groin, by slowing the longshore current, causes sediment deposition upbeach of the groin. Since the longshore current is deprived of its sediment at this point, but immediately reforms to the opposite side of the groin, it removes sediment from this area with no corresponding sediment deposition. The beach is therefore cut back immediately downcoast of the groin (Fig. 8-5). With other structures present, additional groins must be constructed. Once this occurs, the beach will be destabilized immediately downcurrent of each new groin and another groin will be necessary. Consequently, once one groin is constructed, the community is committed to build a series of groins to stabilize the beach that the previous groins destabilized.

FIGURE 8–3

The effects of groins are illustrated by this aerial photograph. Note the sand piled up to the right and the beach erosion to the left of each groin. (*Courtesy Suffolk County Planning Department*)

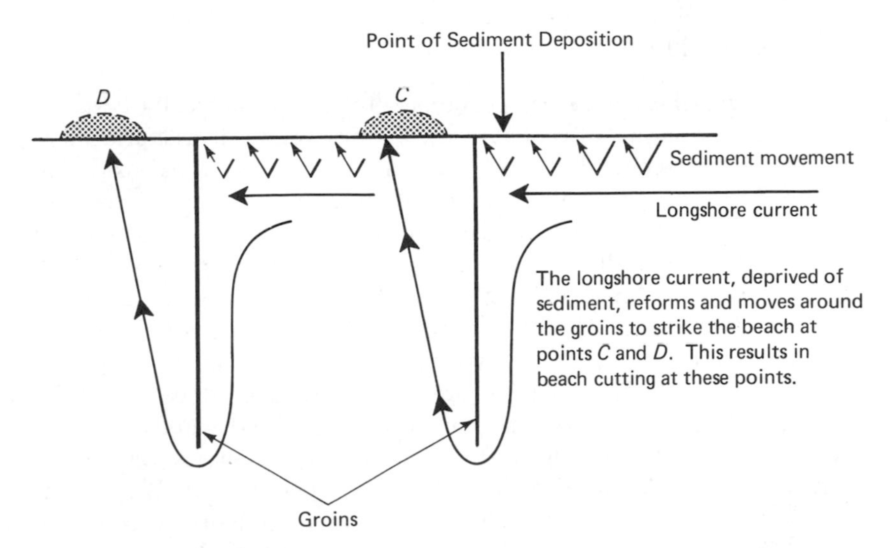

FIGURE 8–4

Schematic illustration of the effect of groins on sediment transport.

FIGURE 8–5

In this area the longshore current moves from right to left. Note the beach building to the right of the jetty (Point *A*), and the beach destruction to the left of the jetty (Point *B*). (*Courtesy Suffolk County Planning Department*)

coastal landfill

Salt marshes are both economically and ecologically vital to coastal and estuarine areas. They play an important role as storm surge buffers, erosion-control devices, sediment traps, and as oxidation basins that serve to break down deleterious materials prior to their entering the estuarine waters. They also act as wildlife habitats and estuarine nurseries. Although estimates vary as to the amount of organic material produced, and thus available as food for estuarine food chains, their value is considerable. For example, it is estimated that Georgia marshes produce 4.4 to 8.9 tons (dry weight) of organic matter per acre per year; Virginia marshes are estimated to produce 3.0 to 7.0 tons (dry weight) per acre per year and Long Island marshes to produce 2.3 to 3.7 tons (dry weight) per year. This compares very favorably to the 1 to 1.5 tons (dry weight) per acre per year yields of wheat production and the 0.33 ton per acre per year produced in the deep ocean. Salt marshes are, therefore, one of the most productive natural resources available in a time of dwindling food resources.

It is calculated that tidal currents transport 45% of this material into adjacent estuarine waters, where it forms the base of complex food chains. Since two thirds of the total commercial fishery catch are estuarine-dependent species, it is estimated that each acre of salt marsh produces a tangible benefit of $78 annually.

In spite of their proved value, thousands of acres of salt marshes are destroyed annually. For example, 25% of the salt marshes in Suffolk County, New York, were destroyed between 1964 and 1971.

The major factors in marsh destruction are landfill operations to create waterfront real estate and deposition of dredge spoil. Regardless of the source, the destruction of marsh lands is an irreversible process that directly reduces the productivity of many coastal areas.

effects on man

The consequences of pollution on the human race are widespread and range from a reduction in the recreational value of beaches and waterways, to unaesthetic drinking water, to serious health hazards. Consequently, any of the practices discussed in the preceding sections will affect man to some degree. For example, any mechanism that serves to reduce the numbers of organisms in a natural food chain affects man not only by reducing the amount of food available to the human population, but also by adversely affecting the economics of the commercial fishing industry. Thus practices as divergent as the ocean disposal of sludge, the increased mercury concentrations in fish, the release of heated effluent, and the construction of groins will all adversely affect the economic, recreational, and/or biological well-being of at least some segment of the human population.

Rather than consider each pollutant in detail, this section will discuss only selected examples of the consequences of pollution on man. The factors to be considered here will be the well-studied metals —lead, mercury, and cadmium—as well as pesticides, nitrate, and carcinogenic compounds found in drinking water supplies. For a complete survey of the impact of pollutants on man, the reader is referred to *Health Effects of Environmental Pollutants* by George L. Waldbott (see Suggested Readings).

metals

The symptoms of lead poisoning include loss of appetite, weakness, and miscarriage, as well as lesions of the neuromuscular system, the circulatory system, the brain, and the gastrointestinal tract. Levels of lead in the blood of Americans range from 0.05 to 0.4 ppm. The threshold levels for acute lead poisoning range from 0.5 to 0.8 ppm. Since lead is absorbed from air, water, and food, it is obvious that many city dwellers exposed to gasoline fumes are presently absorbing lead at or close to toxic levels. Since lead levels in air, water, and food are increasing, it is expected that cases of lead poisoning will show a corresponding increase in the future.

Individuals vary in both their sensitivity and exposure to methyl mercury. For example, populations whose diets consist mainly of fish will have a high exposure to mercury. Symptoms of mercury poisoning include deafness, loss of coordination, insanity, and death. In addition, evidence indicates that methyl mercury may cause genetic damage. As noted previously, pickerel from the Great Lakes have mercury concentrations as high as 5 ppm. Since there have been cases of human mercury poisoning at concentrations as low as 0.2 ppm, this necessitated the authorities in both Michigan and Ohio to ban fishing in not only Lake Erie but also in Lake Saint Clair, the Saint Clair River, and the Detroit River. In addition to the health hazard posed, the economic impact on Ohio's commercial fishing industry was considerable.

Recent U.S. Geological Survey research has found cadmium levels to be above U.S. Public Health Service limits in the water supplies of 12 cities; in addition, cases of fatal cadmium poisoning have been reported in Japan. Cadmium is toxic to all human systems and has been implicated in hypertension and in respiratory and kidney disorders. It also appears to have a genetic effect on experimental animals, inducing malformations in the upper jaws of newborn rats whose mothers had been exposed to cadmium in their food during pregnancy.

Cadmium, mercury, and lead are known to concentrate in food chains, and they are all found in human food sources. It is obvious that exposure can come from airborne as well as waterborne sources; the danger of exposure to these materials is thus increased. The use and input of these materials into the environment should be carefully controlled and curtailed.

pesticides

Pesticides are known to be present in the body fat of human beings. The main source of these materials is through ingestion in food supplies. However, in some cases, pretreated water is known to contain various pesticides. Although all pesticides in use can be fatal in sufficiently large doses, lethal levels cannot be determined adequately, since many variables, such as age, health, sex, and diet, affect the toxicity of these materials.

Since the chlorinated hydrocarbons have only been in widespread use for the past 25 years, possible long-term, sublethal effects, principally carcinogenic and mutagenic in nature, have yet to become evident. It is known that pesticides such as DDT and dieldrin are capable of producing malignant and benign tumors in various organs of experimental animals.

nitrates and nitrites

Increased nitrate levels in water supplies has forced the Public Health Service to establish a nitrate limit of 45 ppm in drinking water. This was considered necessary since concentrations above 66 ppm of nitrates in drinking water can be fatal to infants. At these levels the disease *methemoglobinemia* is induced by intestinal bacteria that have the ability to convert nitrate to nitrite. The nitrite is then absorbed into the bloodstream, where it binds with hemoglobin. This binding action effectively prevents the hemoglobin from accepting the oxygen that is necessary for respiration.

carcinogenic substances

Recent studies indicate that the chlorination of municipal wastewater may be responsible for the addition of chlorinated organic materials to a river system. A 1973 study by the Oak Ridge National Laboratory identified over 50 chlorinated hydrocarbons in domestic sewage effluents. Similar studies in The Netherlands have shown that when contaminated river water is chlorinated prior to distribution to consumers, the chlorination process produced carcinogens in drinking water. The Environmental Protection Agency has also demonstrated that chloroform concentrations ranged from 37 to 152 ppb in five communities that utilize water from the Mississippi or Ohio rivers.

These studies led the Environmental Defense Fund in 1974 to state that public drinking-water supplies are routinely contaminated with carcinogenic substances from the chlorination processes employed at water-treatment plants (pretreatment) prior to the distribution of the water to the consumers. The Environmental Defense Fund concluded that the analysis of tap water from portions of West Virginia, Indiana, Iowa, and Nebraska clearly demonstrated that carcinogens were not removed by the current water-treatment methods. However, these studies do not mean that water-treatment methods are futile. They do indicate that many treatment plants are currently inadequate and must be upgraded.

groundwater

As noted previously, groundwater systems are affected in a completely different manner by the input of various contaminants. Since there is no euphotic zone and, consequently, no food chain as such, many of the materials tend to accumulate in these systems for long periods of time. In addition, if groundwater use is faster than replacement, these systems may become drastically depleted. The two major consequences of groundwater mismanagement are, therefore, contamination and depletion.

contamination

In areas utilizing groundwater, the water is located in aquifers at various depths (Chapter 5). Some aquifers are shallow and the first to be used. Others are deeper and are tapped only when the shallower aquifers become contaminated or depleted. Water drawn up from an aquifer is used and discarded as either household or industrial wastewater. If the wastewater is released into *lagoons* (a common method of industrial waste disposal) or cesspools, it will slowly percolate back into the aquifer under the influence of gravity. In addition, pesticides, fertilizers, and de-icing compounds also tend to enter the water table through percolation. As water trickles through the sediment overlying the water table, it is purified to some degree. Most of the bacteria as well as the suspended organics and most of the pesticides are filtered out or absorbed by the sediment within approximately 30 feet (ft) of the source of input. Soluble substances such as dissolved metals and nutrients pass unchanged into the water table. The nutrients are converted from their organic to their inorganic form by bacterial action. In the inorganic form they are usable by plants but are, in the aquifer, out of the euphotic zone and therefore spatially unavailable to plants for use in metabolic processes. Consequently, large reservoirs of unused nitrate and phosphate tend to build up. The dangers of nitrate contamination were illustrated when the U.S. Public Health Service established an upper limit of 45 ppm on nitrate levels in drinking water. This federally established limit forced a water company in San Joaquin Valley, California, to notify its customers in 1966 that its water was unfit for consumption by infants since it contained a level of 45 ppm of nitrate and thus created a risk of inducing methemoglobinemia. The situation in California occurred because the intensive agricultural activity in this region necessitated extensive chemical fertilization and irrigation within a short time period. This, coupled with naturally high soil nitrate levels, resulted in a rapid increase of nitrate levels in local aquifers.

Excessive phosphorus, in any form, does not have any apparent deleterious physiological effects. In addition to the possible harmful effects of nitrate on infants, it is to be noted that both phosphorus and nitrogen do pose harmful environmental effects.

These effects arise because recharge is generally into shallow aquifers. Because of lateral flow, nutrients eventually reach the euphotic zone by flowing into streams, lakes, and so on. As this occurs, the nutrient levels of surface waters increase and eutrophication follows.

The contamination of aquifers by nonnative heavy metals was illustrated in 1962 on Long Island. In that year, water from private wells was found to contain chromium and cadmium levels hundreds of times above accepted U.S. Public Health Service levels. The source of these metals was traced to a local metal-plating plant that discharged industrial waste water into leaching basins in the early 1940s. Additional investigations revealed a body of contaminated groundwater 4200 ft long and 1000 ft wide moving southward, via lateral flow, into previously uncontaminated wells. The discovery of the contamination of these shallow wells necessitated the tapping of deeper aquifers, fortunately uncontaminated, to supply this area with safe drinking water.

A potential source of contamination is the disposal of waste materials by injection wells. In the mid-1960s federal laws restricting the disposal of liquid wastes into surface waters forced industries to seek other means of waste disposal. Consequently, in many portions of the country, disposal of materials such as oil-well brine and industrial waste is accomplished by injection into deep, permeable rock structures.

This system of liquid waste disposal may present unforeseen environmental hazards. At the present time little is known of the fluid movements and physical and biological changes that occur in deep aquifers under the process of injection. An example of not considering the physical and chemical implications of injecting waste substances occurred in Erie, Pennsylvania when a paper-mill injection well, operating at a depth of 1600 ft, exploded, spewing 50,000 gallons/day of a sulfite liquor waste into Lake Erie. The well blew up when the injection tubing corroded and the build up of pressure within the aquifer forced the waste liquid back up the well casing to the surface.

One of the most widely known cases of unexpected problems associated with subsurface injection was the apparent stimulation of earthquakes between 1962 and 1966 during the operation of injection wells for the disposal of noxious chemical wastes at the Rocky Mountain Arsenal near Denver, Colorado. It was theorized that the waste fluids may have acted to permit slippage of rock masses by lubricating faults deep in the earth's crust.

Less well known but of greater potential hazard was the prediction that these substances would tend to migrate and thus contaminate aquifers as far east as Kansas and Nebraska.

Alarmed at the increasing popularity of this disposal method, the commissioner of the Federal Water Quality Authority in 1970 opposed injection disposal without strict controls and clear demonstration that such disposal will not harm present or potential subsurface water supplies.

depletion

As an area develops and population increases, the volume of wastewater increases, resulting in an increase in the deleterious substances leaching into local aquifers. Eventually the water will become contaminated and reach a state where it will be either unsafe or unaesthetic for use. It is then necessary to exploit a new aquifer at a deeper level than the one initially used and contaminated.

The practice of using, contaminating, and exploiting new aquifers cannot continue indefinitely. Eventually even the deeper aquifers become endangered, either by depletion or contamination. Although recharge of water is not directly into the deeper aquifer, contamination is possible since water is drawn from the deeper aquifer, used, and released as wastewater to the superficial aquifer. Consequently, a situation arises where water is withdrawn but not returned to the same water table. This will cause a reduction in the hydrological head of the deeper aquifer. As the hydrological head is reduced, an increased vertical flow from the shallower aquifer will occur and water will tend to leave this aquifer at a rate faster than it is replaced by rainfall. This causes the contaminated water to move downward at an increasing rate.

In addition to contaminating the deeper aquifer, this increase in vertical flow will tend to deplete the upper aquifer. This is undesirable because it is this aquifer which supplies the majority of water to streams, lakes, and rivers. In many areas the reduction in flow or the actual drying up of streams and lakes is a visual indicator of a lowered water table. A study in the early 1950s in Nassau County, New York, showed that a 10-ft drop in the water table was responsible for a significant decrease in the flow of two streams in the area studied.

Another consequence of excessive groundwater depletion may be subsidence (the sinking of the land surface to fill the void created by removal of the groundwater). The most common form of subsidence occurs in regions where the subsurface rock strata are interspersed with layers of loose, fine-textured sand and silt. As water is removed from the pore spaces this material becomes compacted. In these areas the land surface may sink at rates of ½ inch to 2 ft/year. Severe subsidence of this type has occurred in the Houston–Galveston area, where it is estimated that the Johnson Space Center could sink sufficiently to become flooded. In Mexico City subsidence is responsible for the sinking of many of the older buildings. In this case these buildings have sunk to a point where their first floors are below street level. It is possible to stop subsidence and, in some instances, accomplish a small amount of "rebound" by halting the excessive pumping of groundwater.

A less common type of subsidence occurs in areas of limestone bedrock, in which water has formed cracks and caverns. Removal of the water from these underground reservoirs may permit the upper soil layers to enter the void and form cavities nearer to the surface. In these cases a minor surface shock can trigger a sudden cave-in. Florida and Alabama are particularly susceptible to this type of subsidence.

Once it is obvious that the deeper aquifers are becoming contaminated, most communities begin a sewage-treatment program. The most common type of treatment in the United States is known as secondary treatment (Chapter 9). These systems are incapable of sufficiently purifying wastewater for reuse, and so it must be disposed of either by pumping the partially treated effluent into rivers, bays, or the ocean or, if the facility is not located near a body of water, by lagooning.

If the treated effluent (from secondary treatment) is released into a surface-water system, it will increase the BOD and also lead to eutrophication of that system. Discharging rather than recharging this water will also lower the water table, reduce stream flow, and, in coastal areas, encourage saltwater intrusion. Disposal of this material into lagoons or leaching basins will result in contamination of the aquifers (in a manner similar to cesspool recharge).

SUGGESTED READINGS

Ehrenfeld, David W. 1970. *Biological Conservation.* New York: Holt, Rinehart and Winston, Inc.

Ehrlich, Paul R., and Anne H. Ehrlich. 1972. *Population, Resources, Environment.* San Francisco: W. H. Freeman and Company.

Ettlinger, Ken W. 1974. *Long Island Groundwater: An Environmental View.* New York: Moraine Audubon Society.

Hodges, Laurent. 1973. *Environmental Pollution.* New York: Holt, Rinehart and Winston, Inc.

Kilbourne, E. D., and W. G. Smillie (eds.). 1969. *Human Ecology and Public Health.* New York: Macmillan Publishing Co., Inc.

Wagner, Richard H. 1974. *Environment and Man.* New York: W. W. Norton & Company, Inc.

Waldbott, George L. 1973. *Health Effects of Environmental Pollutants.* St. Louis: The C. V. Mosby Company.

QUESTIONS

1 / List the four categories of pollutants.

2 / Discuss the difficulties involved in categorizing a given contaminant.

3 / Explain the various mechanisms that may lead to oxygen depletion.

4 / What are the environmental effects of oxygen depletion?

5 / Discuss the origin of excessive plant nutrients.

6 / What environmental effects may be expected should increased fertilization of a system occur?

7 / What criteria are necessary in order to consider a substance an agent of biological dysfunction?

8 / Explain biological magnification.

9 / Explain why sediments may be considered a pollutant.

10 / Explain, in light of Chapter 6, how coastal landfill operations may decrease the productivity of the oceans.

CHAPTER 9

water-treatment methods

Prior to the popularity of the flush toilet in the early nineteenth century, the bulk of household wastes were disposed of with minimal water loss, in a dry or semidry state (with the exception of a relatively small amount of liquid released as wash water). These wastes were released into the dry pits of the backyard outhouse or were transported to vacant fields as "night soil" (so named because of the tendency of individuals to dispose of their waste on another's property under the cover of darkness). After the flush toilet became popular, however, large amounts of water were required to remove wastes. This use of clean water, 5 gallons per flush, to transport a minimal amount of dry waste or urine led to a twofold problem: the contamination and consequent depletion of previously potentially usable household water, and a subsequent increase in the volume of contaminated water. The increase in the volume of liquid sewage initially caused serious overflow problems with the cesspools (which replaced the dry pit) in many communities. In order to alleviate the problem of overflow, connections with the open street and storm drains were made. This led to an increase in filth, odors, and epidemics of waterborne disease (cholera, typhoid, infectious hepatitis, etc.), which eventually forced the construction of underground drainage systems or sewers.

Initially these sewers were merely a means of transporting the wastewater to a waterway, where it was released in a totally untreated and raw state. This naturally led to anaerobic conditions in the waterway in the immediate vicinity of the outfall and the elimination of the normal aquatic or marine populations. In municipalities of low population density the input of sewage led only to localized problems (minimal water pollution and the elimination of only local aquatic or marine populations), since the bacteria inhabiting the waterway were able to decompose and thus purify the relatively small amount of wastewater in a reasonably short time. Consequently, downstream or downcoast communities were generally unaffected.

As population continued to increase, so did the amount of sewage generated. Eventually, many communities reached a point where the disposal of this material into adjacent waterways led to widespread pollution of the particular waterway. The problems, at this point, became regional rather than local, and began to affect downriver communities that relied on the river as a source of drinking water as well as for recreational purposes. Coastal communities that disposed of their wastewater by outfalls into marine or estuarine systems generally degraded the water quality of downcoast or downcurrent communities. This led to a decrease in the recreational value of their beaches, as well as the contamination of commercially valuable shellfish areas.

Communities that continued to rely on cesspools for disposal of wastewater were not totally immune to waste-disposal problems. Cesspools release sewage in various stages of decomposition into the groundwater, which, in many cases, serves as a water supply. These communities, in order to obtain suitable water, were forced to construct deeper wells, leading to the problem of drawdown (Chapter 5). This served to bring contaminated water from shallow aquifers down into the aquifer supplying that community's water supply.

Water quality of the adjacent receiving waters eventually declined to such a state that the sewage had to be treated in order to preserve or restore the integrity of the waterway or aquifers. As population continues to increase, however, it will be necessary to construct more highly advanced and expensive treatment systems to deal with the increased sewage generated by an expanding population.

Currently many communities find it necessary not only to treat wastewater prior to its release (posttreatment) but also to treat water prior to use (pretreatment). Posttreatment or wastewater is necessary to maintain healthful, attractive living conditions as well as to maintain the natural biological communities present in those marine and freshwater systems that serve as receiving waters for a community's wastewater. It is also necessary in order to maintain the recreational and commercial value of these systems. Pretreatment is necessary since most posttreatment methods do not purify the wastewater sufficiently for reuse when the water reaches another community that must use it for consumptive purposes. In addition, as the water travels to its point of reuse, many additional contaminants may enter the system. Contaminants that commonly enter a community's water supply on its way to that community are: various agricultural wastes, pesticides, hydrocarbons from road runoff, and sediments.

pretreatment

To be suitable for use, water must be free of pathogens, should have no undesirable tastes, odors, colors, or turbidity (cloudiness), and should not contain harmful or unaesthetic materials. Thus the contaminants that must be removed in pretreatment may affect the physical, chemical, or microbiological characteristics of water.

The physical characteristics of a water supply are said to be affected if the water has an unsuitable taste, color, odor, or turbidity. Tastes and odors in water are due to decomposed organic matter and organic chemicals. Various organic constituents leached from soil or decaying vegetation often impart undesirable colors to water. Color intensity is generally determined by a visual comparison using glass tubes (Nessler tubes) containing different standard colors. Suspended clay and organic materials will increase the turbidity of water. The turbidity depends on both the size and concentration of the particles suspended in the water and is customarily determined by measuring the reduction of light as it passes through a water sample.

The chemical characteristics of a water supply are affected by the presence of dissolved compounds. Chloride ions find their way into groundwater primarily through saltwater intrusion. Chlorides in surface waters originate mainly from various industrial inputs, as well as from salt spread on roadways as de-icing compounds. Chloride concentrations above 250 mg/liter will impart an undesirable taste to drinking-water supplies. Hydrogen sulfide from anaerobic decomposition processes also leads to undesirable tastes and odors.

Waters containing calcium and magnesium ions are said to be hard and react with soap to form scum (this led to the development of detergents to combat this problem—see Chapter 7). There are two types of water hardness—temporary and permanent. Temporary hardness is caused by calcium and magnesium in its bicarbonate form. If this water is boiled, carbon dioxide will be evolved and an insoluable carbonate precipitated. The precipitation of carbonate is the cause of "boiler scale." Permanent hardness is caused by the presence of calcium and magnesium in the form of nitrate, sulfate, or chloride. These constituents cannot be removed by boiling. Two other materials that are commonly found dissolved in water supplies are iron and manganese. These elements are undesirable because they cause decolorization of clothing when it is laundered. In addition, as manganese oxide and iron hydroxide they tend to precipitate and form incrustations in water mains, plumbing, and so on.

Pesticides, because they are only slightly soluble in water, are mainly found in their suspended state and/or coating sediment particles that enter waterways during irrigation operations or from erosion processes during periods of heavy rainfall. Some of the pesticide does, however, dissolve in the water. Pesticides, therefore, affect both the physical (in their suspended state) characteristics and the chemical characteristics (in their dissolved form) of a water supply. Consequently, different pretreatment processes (carbon adsorption to remove the dissolved material and sedimentation, followed by filtration to remove the particulate forms) must be employed to remove the total pesticide load from a given water supply. Dissolved metals of many elements may possibly enter a water supply (generally in their dissolved form). The particular type and concentration of the metal is dependent upon the soil composition as well as the industrial processes of the surrounding area.

The microbiological characteristics of water are affected by the presence of bacteria, viruses, algae, fungi, or protozoans. The bacteria *Escherichia coli* (termed *coliforms*) are found in the intestines of warm-blooded animals. Although they are harmless (nonpathogenic), their presence in a water supply indicates that the water has been contaminated with sewage and that there is a possibility that pathogenic bacteria are also present. Any one of a number of analyses may be employed to determine the presence of coliforms. In general, a series of tests termed the presumptive, confirmed, and completed tests are used in conjunction with and to amplify the MPN method (Chapter 7).

Algae impart taste and odor to water supplies and generally become a problem when the source waters are high in plant nutrients. Since they are plants, and thus require sunlight, they are not a problem in communities utilizing groundwater as a water-supply source. Fungi are plants that are capable of growing without sunlight and are, thus, able to infest water mains, plumbing, and so on. When this occurs, the fungi may produce unpleasant tastes and odors in the water and may even cause clogging of water mains. Protozoans that may be present in water supplies include *Endamoeba histolytica,* which is the cause of amoebic dysentery. Since over 50% of the pathogens in water are known to die within 2 days, and 90% within 7 days, reservoir storage is reasonably effective in eliminating bacterial contamination of water supplies. However, a few pathogens may survive for over 2 years, and hence the water must be disinfected prior to distribution to consumers for human consumption. Disinfection is commonly accomplished by chlorination, which has an immediately lethal effect on bacteria.

In addition to chlorination, which is routinely performed, other methods may be utilized to make the water supply safe and aesthetic. The method selected is totally dependent on the nature of the water that is to be treated. Many of the methods that may be employed in the pretreatment of water are also used in the treatment of domestic and/or industrial wastewater. These overlapping methods will be only mentioned here and covered in greater detail in the sections dealing with these posttreatment methods.

Larger suspended materials can generally be removed from a water supply by holding the water in sedimentation basins, where the larger particles will settle under the influence of gravity. The efficiency of the sedimentation process may be increased by the addition of chemicals to the water. This process is known as *chemical coagulation* and is discussed in conjunction with tertiary treatment methods (p. 147). Finer particles that are not removed by these methods are generally removed by filtering the water through sand, diatomaceous earth, or microscreens. The filtration process is generally preceded by coagulation or sedimentation. This prevents the filters from becoming clogged by an inordinate amount of sediment.

Tastes and odors are removed from water supplies by aeration. In this process the water is sprayed or trickled over cascades. These processes break the water into fine droplets and permit escape of the gases that cause the objectionable odors or tastes. Activated carbon is also commonly used to remove tastes and odors.

Hard water is generally subjected to chemical treatment to remove the materials (calcium and magnesium salts) that serve to cause hard water. The removal of these substances is not required to make the water safe for use, but it is necessary in order to prevent water-main clogging and to preserve household plumbing. In addition, water softening will reduce household soap consumption as well as decrease the need for detergents. Either of two methods is used to decrease water hardness: ion exchange and the lime-soda process. In the lime-soda process, lime [$Ca(OH)_2$] and soda ash ($NaCO_3$) are added to the water. These materials react with and precipitate the calcium and magnesium salts, which are removed from the water as insoluble sediments. Ion-exchange processes are discussed on page 155.

Many areas throughout the world are presently obtaining water by the desalinization of seawater. Although the costs of desalinization are high, it is competitive with the costs involved in obtaining alternative sources of fresh water in many water-short countries. The largest desalinization plant in operation is in Kuwait, where 5 million gallons of fresh water are produced per day. In many areas that utilize groundwater as a water source, saltwater intrusion is a definite problem. Many of these communities are presently considering desalinization as an alternative water source. Water may be desalinized by distillation, demineralization, electrodialysis, and reverse osmosis.

The distillation of seawater has been used for many years as a means of obtaining fresh water. Currently, it costs approximately $100 to produce 1 million gallons of fresh water. It is estimated that costs can be reduced by approximately 60% by constructing electric generating plants. These plants would be capable of generating steam for electric power as well as to distill seawater. Demineralization, electrodialysis, and reverse osmosis will be discussed in conjunction with industrial water-treatment methods.

As noted previously, the pretreatment required depends solely on the characteristics of the water source. Groundwater generally is not turbid and thus need not be subjected to the various sedimentation processes. Most groundwaters are hard, however, and contain large amounts of iron and/or manganese. Consequently, water softening and the removal of these two elements is often necessary.

River water, since it is generally turbid, often requires chemical coagulation and filtration prior to use. In addition, industrial development along a water course may require the removal of many compounds that may enter a system from these sources. A waterway that passes through agricultural areas must generally be treated to remove pesticides prior to distribution to the consumer. In addition, most, if not all, waters utilized today should be disinfected prior to use.

domestic posttreatment methods

Most conventional sewage treatment plants are designed to merely remove or reduce the organic materials and thus decrease the BOD of the waste water. In addition, the water is generally chlorinated prior to release to remove any bacteria that may be present. In most conventional treatment plants, little attempt is made to remove the phosphorus or nitrogen present in the influent. There are three basic types of treatment that may be used in treating domestic sewage: primary, secondary, and tertiary. These terms refer to the quality of effluent released rather than the actual treatment process employed. For example, primary treatment produces the lowest-quality effluent in the minimal time period, whereas tertiary-treatment systems produce the highest quality. The processes employed in the various types of tertiary treatment may vary widely, depending on the type of tertiary plant constructed. In areas of low population density, primary treatment may be sufficient, since it will perform the initial treatment stages and the waterway accepting this minimally treated effluent will further purify the water subsequent to reuse downstream. Problems arise when the population grows to a point where large amounts of sewage are treated minimally and then released. When this occurs, the bacteria of the river cannot purify the wastewater sufficiently prior to reuse downstream. Consequently, a more sophisticated plant must be constructed to achieve a higher degree of wastewater purity prior to release into a given waterway.

Primary treatment is primarily a mechanical process which involves grit removal, grinding, flocculation, sedimentation, and skimming. The grit-removal chamber and screening systems are designed to remove large suspended and floating materials, as well as sand and grit, before the influent passes into the actual treatment plant. This is considered necessary since these materials may interfere with the machinery and equipment in the remainder of the treatment area. Grinding is also generally accomplished outside the actual plant and makes use of comminutors to physically break solids down to a size of 6 mm. In flocculation the wastewater is agitated by mechanical stirring. This tends to cause the small suspended solids to collide, forming larger particles (*flocs*) that will then settle out. In the sedimentation stage, the removal of additional suspended solids is accomplished by gravitational settling (Fig. 9-1). Generally, the actual time of wastewater treatment in this type of system is 6 hours. The longer the wastewater is held in the settling stage, the greater the removal of settleable solids. Obviously, the more individuals that release sewage into this treatment system at a given time, the less time the material can be held and the lower the efficiency of the system. A well-run, efficient primary-treatment plant will typically remove 60% of the suspended solids, 35% of the BOD, and 30% of the COD. In addition, 20% of the total nitrogen (as suspended solids) and 10% of the total phosphorus (as suspended solids) are removed. It is to be noted, however, that none of the material in its dissolved form is removed. Prior to release, the effluent is generally chlorinated (Fig. 9-2).

Secondary treatment involves the use of biological methods (trickling filters and/or activated sludge) in addition to the usual methods employed in primary treatment. These biological methods approximate the natural purification processes that wastewater would be subjected to in a natural waterway in a shorter time period and reduced amount of physical space. The activated-sludge process involves the circulation of bacteria through organic waste in the presence of air or oxygen which is continually supplied. The process

FIGURE 9–1

Settling tanks—primary treatment plant. (*Photo: G. Marquardt*)

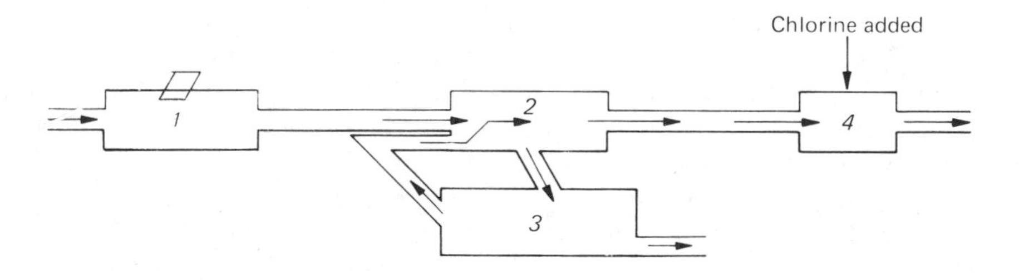

FIGURE 9–2

Schematic diagram of a primary treatment plant. Raw sewage enters the first tank, and the solids on the surface are removed by a skimmer. In tank *2* the heavy materials are allowed to settle. These solids go into a sludge tank (*3*). From tank *3* the sludge is removed and the liquid component is recycled back to the settling tank (*2*). After the settleable solids have been removed, the liquid goes into a contact tank (*4*) where chlorine is added prior to disposal into rivers, etc.

generally operates as follows. Effluent from primary settling tanks is diverted into another, aerobic tank, and is further treated (the secondary stage) by microorganisms originating from the settled sludge. These suspended and dissolved organic wastes are aerated, mixed, and thus undergo adsorption, flocculation, oxidation, and general biological decomposition prior to release into a sedimentation tank (Fig. 9-3). In the sedimentation tank the sludge, with its large concentration of microorganisms, is allowed to settle out. The liquid effluent is released, after chlorination, either into lagoons to allow for groundwater recharge or into a waterway. A portion of the settled sludge is retained to seed the next load of wastes from the primary tanks. This process serves two purposes. It clarifies the wastewater by adsorbing a large percentage of the colloidal and suspended solids on the surfaces of the sludge particles, and it serves to oxidize the organic material entering the system. This type of treatment is analogous to the natural processes that would occur in a river. Initially, wastewater would be subjected to anaerobic conditions (primary treatment), and as it moved farther downstream aerobic conditions would prevail (secondary treatment—the addition of air). The processes in this type of plant occur in a much shorter distance, and the comparatively clean water that is added to the river, although it cannot be used immediately for human consumption, will be purified sufficiently by the river's natural processes to be used by the next community downstream. This is assuming that the plant is not operating above its designed capacity. In a well-run plant, much of the initial stress and impact of the sewage is removed from the natural waterway by the treatment plant.

In the trickling filter method, after the removal of sludge in set-

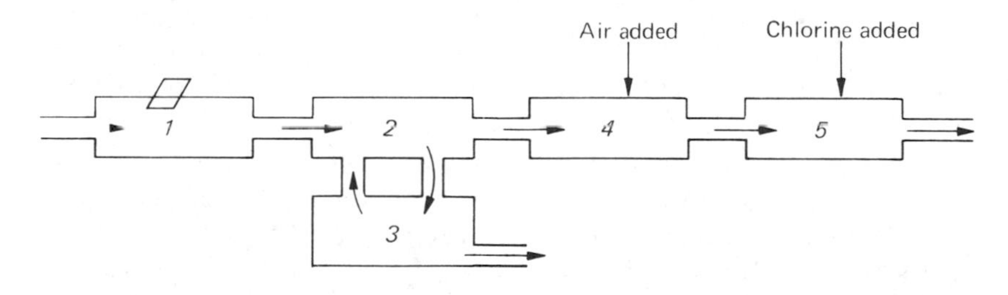

FIGURE 9–3

Schematic diagram of a secondary treatment plant. Steps *1*, *2*, and *3* proceed similarly to those in a primary plant. In step *4* air, oxygen, or ozone is added to aid in aerobic decomposition. Following aeration, chlorine is added (5) prior to disposal.

tling tanks, wastewater is sprayed over and allowed to percolate through beds of crushed stone coated with "biological slimes" (various growths of bacteria, other microorganisms, and algae). As the material is sprayed and trickles through the stone beds, it is brought into contact with, and adsorbed by, the surface of the slimes. These organic materials are then partially decomposed by the bacteria and fungi component of the slime. Since the slimes require phosphate and nitrate for their metabolic processes, there is a minimal removal (5%) of dissolved phosphate and nitrate in this process. As expected, the slimes grow readily under these conditions, eventually slough off, and may be removed. New slimes grow rapidly and soon recoat the stones; thus the process continues with little or no decrease in efficiency. Near the surface of the trickling filter, aerobic decomposition occurs, whereas farther down, where less atmospheric oxygen diffuses, the decomposition becomes anaerobic. This process is the reverse of the activated-sludge method discussed previously. A well-run secondary-treatment plant will typically remove 90% of the BOD, 80% of the COD, and 90% of the suspended solids, including 50% of the total nitrogen and 30% of the total phosphorus. Approximately 5% of the dissolved materials is also generally removed. Both primary- and secondary-treatment plants are designed primarily to reduce the BOD and suspended solids of the entering waters, and thus release a "cleaner" end product into the surface waters. The surface waters will then presumably be able to perform the final purification steps prior to reuse. These plants are not designed to remove heavy metals, viruses, hormones, or other exotic chemicals. Thus, as a community grows and industry, hospitals, and so on, begin adding waste to these plants, less and less of the total waste load is removed and problems arise.

In tertiary treatment (Fig. 9-4) the sewage is subjected to one or more additional processes, either prior to or after receiving conventional primary and secondary treatment. This additional treatment is an attempt to solve particular contaminant problems specific for a given area, such as the removal of certain industrial chemicals, dissolved metals, phosphates, or nitrates. A tertiary plant is designed specifically for a given community, since it would make little economic sense to add a tertiary phase that would remove heavy metals from wastewater in a community that does not have nor intend to have industry of this type. Tertiary treatment will, therefore, vary depending upon the actual wastewater generated or anticipated to be generated in a particular area. In the following discussion only those tertiary methods used to treat domestic wastes will be considered. The treatment of wastewater containing industrial wastes, trace metals, and so on, will be considered under industrial wastewater treatment.

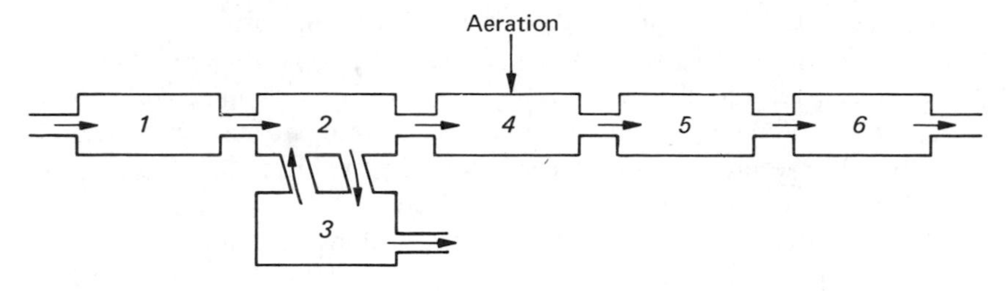

FIGURE 9–4

Schematic diagram of an advanced waste treatment facility. Steps *1* through *4* proceed simliarly to those in a secondary plant. Wastewater would then pass into a lagoon (*5*) containing rooted aquatic plants. Due to the high nutrient levels in the wastewater, the plants, chosen for food value for livestock, will grow rapidly. Succeeding lagoons (*6*, etc.) contain plankton and secondary and tertiary consumers.

Common methods employed in the tertiary treatment of domestic wastes are chemical coagulation and filtration, carbon adsorption, chemical oxidation, ammonia stripping, and advanced biological treatment.

In chemical coagulation and filtration, the wastewater is subjected to an additional treatment stage by the addition of chemical coagulants that react with the suspended material to form flocs that will then settle out under the influence of gravity. If phosphate removal is required, it can be accomplished by the addition of lime, which causes the phosphate to precipitate. In addition to lime, many other chemical coagulants are available. The most commonly used are aluminum sulfate (alum), ferric sulfate, and ferric chloride. Recently a number of synthetic water-soluble polymers have also been developed for use as chemical coagulants. After coagulation the wastewater is filtered through sand, diatomaceous earth, or microscreens prior to chlorination and release. This process can also be used in the purification of poor-quality water at the purifying station prior to its distribution to consumers.

Carbon adsorption involves the passage of wastewater through columns packed with activated carbon. The carbon has many active sites (positively and negatively charged regions) on the surface of each granule. When wastewater with its load of dissolved ions passes through the column, it will come in contact with these charged sites, form bonds, and adsorb on the carbon. This process effectively removes practically all phosphate ($PO_4^{\equiv}$), as well as any dissolved metals present as ions. It fails to remove many forms of nitrogenous compounds. Consequently, wastewater treated by this process will remove nearly all the dissolved materials except various nitrogenous compounds. When this effluent is released into surface waters, especially into the marine or estuary environment, nitrogen/phosphorus imbalances will occur (see Chapter 4 for details).

An additional problem with this type of treatment arises when the active sites on the carbon become "saturated" with the impurities removed from the wastewater. When this occurs, the carbon must be regenerated by heating to 925°C in an air–steam atmosphere. This process burns off the adsorbed material, which is then released into the atmosphere. Thus, by this process the community trades water pollution for air pollution. If the community decides not to reuse the carbon, it is generally removed and is disposed of in the same manner as is the sludge from the primary and secondary settling tanks. In other words, it becomes a part of the community's solid waste.

In chemical oxidation wastewater is treated by the addition of strong oxidizing agents, such as hydrogen peroxide (H_2O_2) or ozone (O_3), to produce available oxygen for the decomposition of organic wastes. Although the efficiency of removal of organic contaminants is high in this process, the removal of dissolved inorganics is low. This tends to be a definite disadvantage, since the concentrations of inorganic materials will continue to increase in any type of water-reuse cycle. The problem arises due to the fact that in the reuse cycle, additional inorganics will be added to water from which the original inorganics have not been removed by water treatment. After only a few reuse cycles, the concentration of these inorganic substances can increase significantly. Methods such as ion exchange, reverse osmosis, and electrodialysis effectively remove these inorganics from wastewater. These methods are discussed under industrial wastewater treatment.

In ammonia stripping the pH of the wastewater is raised by the addition of lime, and the solution is vigorously agitated with air. This process drives the ammonia from the dissolved liquid state into its gaseous form, where it is released into the atmosphere. Unless methods are employed to trap the ammonia, the air quality in the vicinity of these plants will decline.

Advanced biological treatment systems (biological tertiary treatment) closely approximate the treatment systems encountered in all natural surface waters. At the point of input into a natural system, the wastes are decomposed by anaerobic processes (analogous to primary treatment.) Farther downstream (or away from the initial area of input) aerobic conditions prevail (analogous to secondary treatment). Eventually sufficient oxygen is available and added to the phosphorus and nitrogen, which allows for their ultimate "regeneration" as o-PO_4 and NO_3. These materials, in this form, are now chemically available to the plants inhabiting the system, and are removed by the plants and incorporated into their tissue as organic phosphorus and nitrogen. This effectively removes them from the water column.

At present, Michigan State University is constructing a series of lagoons to treat and recycle wastewater from the city of East Lansing. Effluent from the city is initially treated by the activated-sludge method. After this treatment it is then passed into a lagoon 2 m in depth that contains rooted aquatic plants. These plants were specifically chosen for their food value for livestock feed. Succeeding lagoons will contain phytoplankton and, as a secondary consumer, fish that can be used as a human food source.

A tertiary-treatment plant with an alternate design is presently in operation at the Suffolk County Center in Hauppauge, New York. In this plant the sewage is initially subjected to the preliminary grinding. Chemical coagulation may also be performed if phosphate removal is required; however, this step is not necessary in most cases if the effluent is to be filtered through soil, since this filtration will adequately remove the phosphorus. After the preliminary steps have been accomplished, the sewage enters an aeration tank, where ammonia and nitrite are converted to nitrate by nitrifying bacteria. After nitrification, the material is diverted to a settling tank to remove settleable solids as sludge. The sludge is then removed and recycled back to the aeration tank for further bacterial decomposition. This recycling process serves to reduce the total volume of sludge. The liquid waste is then sent to an anaerobic denitrification tank, where anaerobic denitrifying bacteria, seeking an oxygen source, remove the oxygen atoms from the nitrate and convert the nitrate to nitrogen gas, which is released into the atmosphere. This material is then subjected to an additional settling phase. The sludge settled out at this stage is recycled back to the original aeration tank for further bacterial decomposition, further reducing its volume.

By recycling the sludge in this manner, the total solids that must be removed (as solid waste) are reduced considerably. The effluent, after denitrification, has a total nitrogen concentration of 2 mg/liter. It is then subjected to sand filtration, chlorinated, and released into an adjacent river. This system, therefore, effectively removes phosphorus and nitrogen and reduces the volume of the solid waste. The nitrogen gas that is released into the atmosphere during the denitrification phase presents no problem, since nitrogen comprises 78% of the atmosphere.

In coastal areas, such as Long Island, tentative proposals involve using secondary effluent mixed with seawater for the raising of marine phytoplankton. The phytoplankton would remove the dissolved phosphate and nitrate from the water and grow and reproduce rapidly. They would then be "harvested," disinfected, and used as a food source for shellfish, which are currently cultured and marketed commercially. In these times of increasing population, decreasing food supplies, and increasing costs of livestock feed, biological tertiary-treatment plants are becoming economically feasible and should be considered by every community that is contemplating construction or upgrading of current treatment systems.

sludge disposal

All the treatment systems discussed generate significant amounts of sludge in their settling processes. For example, the Metropolitan Sanitary District in Chicago receives approximately 7 million cubic meters of wastewater daily. After treatment of this wastewater, approximately 1000 metric tons of solids remains and must be disposed of. Consequently, the handling and disposal of sludge is an important part of wastewater treatment, for, if improperly handled, it can lead to additional water-treatment problems as well as health hazards and/or air pollution. Common methods employed in sludge disposal include concentration of the sludge, which is generally followed by digestion, dewatering, incineration, ocean dumping (where feasible and not legally prohibited), composting, or disposal into a "sanitary" landfill.

In the concentration process the sludge is further concentrated by additional gravitational settling. It is then usually digested by exposing it to temperatures of 35°C for extended periods of time. At these temperatures the organic materials are eventually reduced to gases such as methane (65%) and carbon dioxide (30%). In addition, small amounts of ammonia, hydrogen, nitrogen, and hydrogen sulfide are evolved. The methane can be trapped and used as fuel to keep the digestor heated. After approximately 60 days of digestion the sludge is converted into a stable humic material that can be used as fertilizer. The city of Milwaukee treats its sludge in this manner and markets the end product under the trade name Milorganite.

Dewatering (the removal of the water used to flush away the original waste) is accomplished by drying on sand beds. In this process the sludge is merely spread out on a porous sand substrate. A portion of the water evaporates, the remainder filters through the sand. This is a rather poor method, especially in communities that rely on groundwater for consumptive purposes since, in most cases, the sand filtration is inadequate and fails to remove dissolved nitrate, nitrite, and many of the inorganic or organic substances that may be present (see Chapter 4). In addition, the groundwater eventually reaches the surface to feed streams and the like (Chapter 3). When this occurs, the nitrogen levels of the surface waters will rise, leading to undesirable overfertilization and/or nutrient imbalances of these waters. If other chemicals (organics and inorganics) are also present in the sludge, they, too, may eventually enter surface-water systems by this pathway. The disposal of sludge in sanitary landfills leads to many of the same problems encountered in dewatering.

Commonly, dewatering is followed by incineration of the dried sludge at high temperatures. Generally, the heat generated in the decomposition of the dried material is sufficient to maintain combustion. This method is also poor, since incineration usually leads to air pollution.

Many coastal cities dispose of their sludge by ocean dumping (Chapter 8). The New York–New Jersey metropolitan area disposes of 200 tons (dry weight) of sludge per day by this method. Preliminary data indicate that this material may be slowly moving northward toward the beaches of Long Island under the influence of strong ocean currents.

The most intelligent alternative to drying, incineration, landfill disposal, and ocean dumping is composting. This method can be applied to the disposal of garbage as well as sludge and permits the recovery of the organic matter in the form of plant nutrients and humus. In this process the material is placed in a large pile, moistened, and turned periodically and allowed to decompose (ferment) for approximately 6 months. The fermentation occurs at approximately 50 to 90°C, which is a temperature sufficient to kill pathogens. This process permits the recovery of organic material in the form of plant nutrients. After fermentation, the material may be used as a soil builder as well as an organic fertilizer, in much the same way as Milorganite is used. If the composting plant is built in the proper manner, the water released, as well as that used to moisten the material, may be collected and used in the irrigation of crops. Thus, in a properly run composting operation, no contamination of groundwater need occur.

alternative domestic treatment methods

There are three alternative methods of domestic waste treatment that are pertinent to this discussion: the Clivus–Multrum system, the Activated–Sludge Package Units, and meadow–marsh–pond methods. Both the Clivus–Multrum and the Activated–Sludge Package Units are individual home-treatment units, while meadow–marsh–pond methods are, under certain circumstances, capable of treating the wastes generated by entire communities.

The *Clivus–Multrum system* does not use water for the transportation of wastes. The system consists of a toilet and garbage chute, a decomposition chamber located in the basement, and vapor-exhaust ducts. Vegetable wastes are emptied into the garbage chute (located in the kitchen), while human wastes (both solid and liquid) are released in the toilet chute (located in and taking the place of a flush toilet). Organic wastes such as wood ashes, disposable diapers, and household paper may be placed in either chute. These wastes accumulate in the chamber, which is a fiberglass–polyester tank. This decomposition chamber operates similarly to a composter, in which natural decomposition processes are encouraged. As decomposition continues, the wastes break down into smaller and smaller particles, and (since the chamber is inclined) they move toward the bottom under the influence of gravity through a series of screens. Since the material has been screened as it passes slowly through the various decomposition stages, the end product is a fine-textured, odorless humus. Approximately three to six buckets of this humus are formed per year by a family of four. When removed, it may be used as a soil conditioner and fertilizer. The Clivus–Multrum system is said to be completely odor-free; it works with a natural updraft similar to a fireplace. This updraft is created as a result of the decomposition process, which produces heat, causing an upward air movement. The heat generated, however, is too low to foster bacterial growth. Thus pathogens present in the decomposing material are eliminated prior to its final conversion into humus, and the end product is capable of being used on food crops. Household water (bath and dishwater) is not placed into this system but must be disposed of in dry wells. This system therefore reduces the amount of water used in and contaminated by the flushing of toilets, and prevents costly treatment of this water. The water that is used in the household for washing and the like is generally not as badly contaminated, since no sewage is placed into it. Natural filtration through the subsoil from the dry well is generally sufficient to remove the contaminants.

Activated–Sludge Package Units are designed for installation into individual homes. These units are wet systems designed to accept the conventional waterborne wastes. The units presently marketed have minor structural variation, but the basic design generally consists of a multichambered decomposition unit located underground in the backyard. The wastewater first enters an anaerobic settling chamber in which the solids and heavier suspended particles settle out. From this chamber the wastewater enters an aerobic chamber into which air is bubbled. The air fosters the growth of aerobic bacteria and also serves to agitate the wastewater. The agitation causes the smaller suspended particles to collide, forming settleable flocs. Thus the aeration process serves to remove additional suspended particles. After aeration the wastewater is pumped through a series of subsurface pipes and can be used to irrigate lawns and gardens. During the growing season it is assumed that the vegetation so irrigated will tend to remove the dissolved nutrients quite efficiently. During the winter months it is probable that the efficiency of the system will decrease markedly. Since this system is currently under evaluation, only tentative, assumed efficiencies can be given at this time.

The *meadow–marsh–pond method,* as mentioned above, is a technique used to treat the wastewater from entire communities. In this method wastewater is allowed to flow over meadows planted with reed canary grass into a shallow freshwater cattail marsh and finally into a freshwater pond. The marsh and pond are maintained at a constant level by an overflow pipe from which the water is released to groundwater recharge. As the effluent passes through the various components in this system, the vegetation removes the dissolved contaminants; thus, the system discharges potable water as defined by the U.S. Public Health Service and Environmental Protection Agency drinking water standards. At present an experimental meadow–marsh–pond system at Brookhaven National Laboratory on Long Island is treating 50,000 gallons of sewage per acre per day by this method.

industrial waste treatment

A consequence of the population increase in any given area is not only an increase in the amount of household waste generated, but also an increase in the industrial development of that area, as well as a need for increased agricultural production to feed the increasing population. Thus, as domestic waste increases, so does industrial and agricultural waste. A direct result of this population increase is an increase in the contamination of the streams, lakes, and rivers, as well as the groundwaters and coastal and estuarine areas by industrial, agricultural, and domestic wastes.

Most industrial and agricultural waste is currently either released directly into any convenient waterway or is sent to municipal treatment plants where it is mixed with, and given the same treatment as, the community's domestic waste. In either case problems are encountered, since, if it is released into a waterway untreated, it will not only contaminate and thus impair the natural system, but it will also threaten the water quality when the water is withdrawn for consumptive purposes by a downstream community, thus necessitating additional pretreatment. In marine systems it will decrease the recreational value of the water as well as contaminate fish, shellfish, and other organisms. If this waste is treated along with and by the same processes as domestic waste, problems will also arise, since most treatment plants are, as mentioned previously, only designed to remove or reduce organic materials. In addition, many of the chemicals released in industrial waste are toxic to the bacteria vital to the operation of secondary and tertiary plants and will thus seriously interfere with the efficient operation of the plant.

Chemical waste may also be commonly disposed of by placing this material into large metal drums and sinking them in the ocean. Eventually the drums will corrode and this material will enter the marine environment. Another poor method employed is disposal by deep-well injection, in which industrial waste is injected into deep, permeable rock structures (see Chapter 8).

Conventional methods are suitable for the treatment of some agricultural and industrial wastes, however. For example, biological treatment is perhaps the best method for the treatment of wastes generated by the dairy and food-processing industries since their wastes are organic in nature. Wastes from feedlots could be handled by a well-run composting treatment system or mixed with waste and treated by conventional biological methods.

Since the chemical industry produces wastewater that is of a predictable, uniform quality, and the exact pollutants are, therefore, known, it is possible and technologically feasible to chose the proper treatment for the removal of the particular pollutant or pollutants generated. The failure of industry to provide proper water treatment is largely economic rather than due to a lack of the available technology. In other words, it is more profitable to release poor-quality wastewater than to expend the funds for the construction of proper treatment facilities. Processes presently available for the treatment of industrial wastewater include carbon adsorption, ion exchange, electrodialysis, and reverse osmosis.

Carbon adsorption is well suited for the treatment of wastewater from plating and similar industries. Since particular, known metals would be removed on the activated carbon, these metals could be easily recovered when the carbon is regenerated and the metals released from the granules. Carbon adsorption is also used as a pretreatment method to remove pesticides.

Ion exchange is a process whereby wastewater is percolated through various positively and negatively charged resins. The specific resin used is dependent on the material to be removed. A cation-exchange resin exchanges hydrogen ions (H^+) for metallic cations (Pb^-, etc.); as the solution passes through the resin, anion-exchange resins exchange hydroxyl groups (OH^-) for anions in the wastewater. The resins are easily regenerated and the materials when released in the regeneration processes can be recovered.

Electrodialysis consists of the placing of an electrical potential difference across the water to be treated, thus producing an electric current. The cations migrate toward the oppositely charged cathode and the anions toward the anode. Membranes permeable to only anions or cations control the migration of ions and allow demineralized water to be removed at the appropriate chambers (see Fig. 9-5). This method is not suitable for the treatment of highly organic wastes, since organics collect on and clog the membrane. It is an excellent method of treatment for industrial wastewater primarily containing inorganics in water-reuse cycles.

Reverse osmosis is a membrane process that concentrates impurities in a portion of the solution and consequently purifies the other portion of the solution. In reverse osmosis pressure is applied to the more concentrated (contaminated) solution. The pressure forces

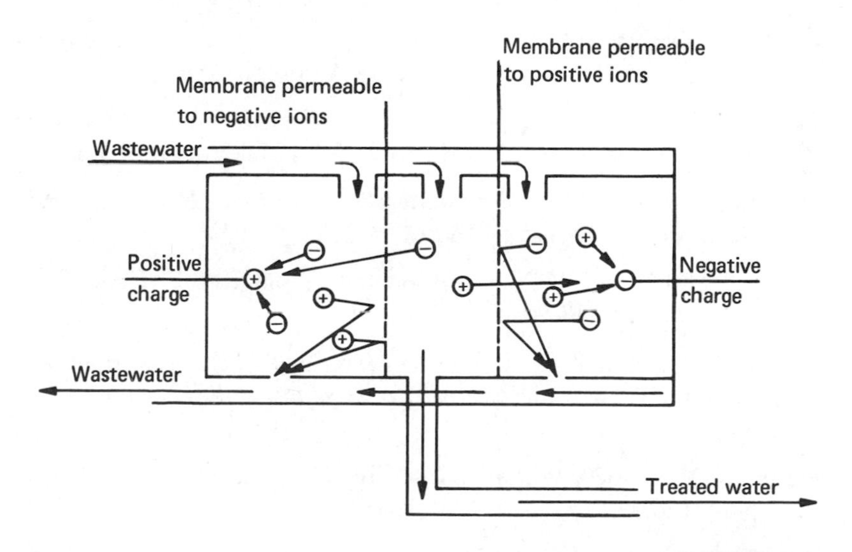

FIGURE 9–5

Schematic diagram of electrodialysis.

the water, but not the impurities, through the membrane (Fig. 9-6) and thereby reduces both the organic and inorganic concentration of the wastewater.

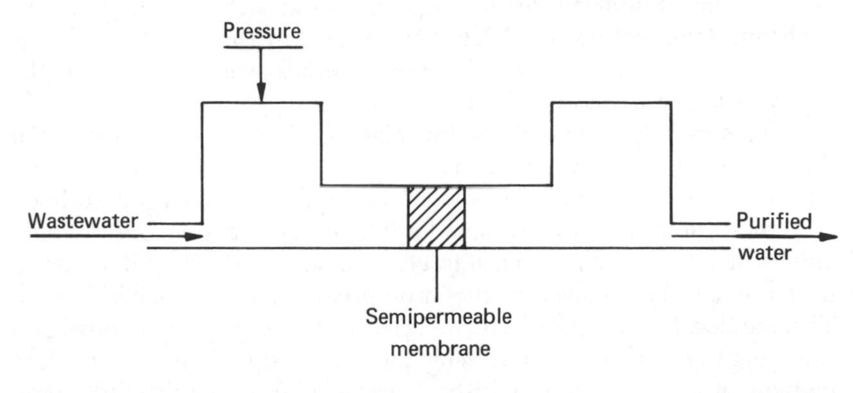

FIGURE 9–6

Schematic diagram of reverse osmosis.

SUGGESTED READINGS

Ettlinger, Ken W. 1974. *Long Island Ground Water: An Environmental View.* New York: Moraine Audubon Society.

Fair, G. M., and J. C. Geyer. 1958. *Elements of Water Supply and Waste Water Disposal.* New York: John Wiley & Sons, Inc.

Hodges, Laurent. 1973. *Environmental Pollution.* New York: Holt, Rinehart and Winston, Inc.

Knapp, Carol E. 1971. "Recycling Sewage Biologically." *Environ. Sci. Tech. 5:* 112–113.

Stander, C. J., and L. R. J. Van Vuuren. 1969. "The Reclamation of Potable Water from Wastes." *Jour. Water Poll. Control Fed. 41:* 355–367.

Stephens, J. H. 1967. *Water and Waste.* London: Macmillan & Company Ltd.

Warren, Charles E. 1971. *Biology and Water Pollution Control.* Philadelphia: W. B. Saunders Company.

QUESTIONS

1 / Relate population growth to wastewater generation in terms of sewage generated, livestock production, and cropland management.

2 / Discuss the differences in pre- and posttreatment methods.

3 / Discuss the differences and similarities among primary, secondary, and advanced waste-treatment methods.

4 / List the problems encountered in solid-waste disposal.

5 / Explain why industrial wastes should be relatively easy to control and treat.

CHAPTER 10

legal aspects of water pollution control

A variety of federal, state, and local laws have been enacted in an attempt to curb the degradation of waterways and drinking-water supplies. In addition, many different water-quality standards have not only been established but are constantly under revision as new data, knowledge, and information become available. This chapter briefly discusses the three major federal acts that govern water quality—the Federal Water Pollution Control Act (FWPCA), the Marine Pollution, Research and Sanctuaries Act of 1972 (the Ocean Dumping Act), and the Safe Drinking Water Act of 1974. The FWPCA and the Ocean Dumping Act attempt to control the water quality of freshwater and marine systems, while the Safe Drinking Water Act sets interim drinking-water standards. Since the water-quality standards set by the FWPCA and the Ocean Dumping Act are often revised, no attempt is made to list absolute standards currently in effect; rather, the reader is referred to the *Federal Register*, which periodically lists the current changes in these standards. The Safe Drinking Water Act, on the other hand, was enacted to establish absolute, although interim, standards. Consequently, the standards set by this act are listed.

federal water pollution control act (FWPCA)

This act was established in 1948 but was placed in its present form through a series of amendments passed in 1972. The FWPCA, in its present form, attempts to accomplish three basic tasks: the regulation of pollution from "point sources," from oil spills, and from hazardous substances. In addition, the FWPCA provides financial aid for the construction of sewage-treatment plants.

A *point source* is considered to be "any discernible, confined and discrete conveyance . . . from which pollutants are or may be discharged." Thus point sources would include industrial plants, municipal treatment plants, and feedlots. A nonpoint source, on the other hand, would be considered to be the discharge of materials from diverse, poorly defined, and scattered sources. Typical nonpoint sources would include cesspool or septic-tank leachate from suburban homes, pesticide and fertilizer runoff from both suburban and agricultural lands, and runoff from roadways and parking lots. (The FWPCA virtually omits nonpoint sources; however, Section 208 of the act does contain a few provisions for the control of nonpoint sources and is discussed on p. 161).

The FWPCA attempts to regulate two classes of point-source discharge: those discharging directly into navigable waters (termed direct dischargers) and those discharging material into publicly owned treatment plants.

direct dischargers

Direct dischargers are subject to both effluent standards and water-quality standards. The effluent standards regulate the amount of a pollutant that may be discharged. In this respect the Environmental Protection Agency (EPA) must set standards for toxic pollutants, which are defined as any material that is toxic to any organism. Toxicity includes "disease, behavioral abnormalities, cancer, genetic mutations, physiological malfunctions, or physical deformations" as well as death. According to this section of the act, EPA will establish a list of toxic materials, and that list must take into account the toxicity, persistence, and degradability of the substances. In addition, the actual or potential presence of the organisms that may be affected and their ecological and economic importance, and the nature and effect of the toxic substances on the organisms, must also be considered.

The limitations placed upon the addition of the substances into navigable waters are generally based on the available pollution-control technology. The FWPCA requires existing dischargers (with the exception of publicly owned treatment facilities) to adopt the best practicable control technology available by July 1, 1977, and the best technology economically available by July 1, 1983.

The Environmental Protection Agency describes but does not mandate those technologies that will meet the required standards. In reality, however, it can be expected that the technology described by the EPA will be adopted by the discharger. This is based on the assumption that an industry that fails to meet a given standard after adopting a technology described by the EPA will be in a better position than an industry that adopts an alternative technology and then fails to meet a given standard.

The water-quality standards are based on the minimum requirements necessary to sustain various water uses, such as recreation and the public water supply. These standards are adopted by the individual states and submitted to the EPA for approval. If the standards are not approved, the EPA may then mandate standards for a given state. It is to be noted that these standards will vary as a result of the uses assigned to a particular water body, as well as to the preferences of the individual states. The EPA has, however, attempted, through its final approval authority, to impose a degree of uniformity on water-quality standards.

As noted previously, publicly owned treatment facilities are not required to meet the effluent standards discussed above. However, those plants in existence prior to July 1, 1977, or approved for federal financing prior to June 30, 1974, must adopt secondary treatment (Chapter 9). In addition, all publicly owned treatment plants must adopt the best practicable waste-treatment technology by July 1, 1983.

According to the FWPCA, any discharger must conform to either the effluent standards or the water-quality standards—whichever is more stringent. To enforce compliance, the Environmental Protection Agency has chosen a permit system. Permits are issued that define the maximum levels of discharge permissible for compliance with all pertinent standards.

discharger to publicly owned treatment facilities

The EPA maintains control over these types of discharge by the establishment of pretreatment standards. These standards are designed to prevent the discharge of any material through a public treatment facility that would interfere with, pass through, or be incompatible with that facility.

Since these treatment facilities are designed to remove BOD, suspended solids, fecal coliforms, and the like, pretreatment may not be required for compatible wastes. The operation of the treatment facility must, however, be protected from incompatible wastes that would interfere with the normal functions of the plant and/or materials that are not susceptible to treatment. These materials would tend to pass through the facility and have an adverse effect on the receiving waters.

Although the FWPCA does not require a discharger to obtain a permit, permits may be required by the municipality. In addition, the treatment facility must obtain a permit for the discharge of the treated wastes and must inform the EPA or appropriate state agency (depending on which agency issued the permit) of any substantial change in the volume or character of the materials entering the treatment facility. In the case of violations, the EPA or the state may restrict or prohibit any new "tie-ins" to the treatment system.

oil spills and hazardous substances

The release of hazardous substances and oil spills are subject to a different regulatory process. In the case of hazardous substances, the Environmental Protection Agency is required to designate a list of these substances. It was, felt, however, that the term "hazardous substance" was poorly defined and could not be regulated along with oil; consequently, a list of hazardous substances has yet to be designated.

The facility responsible for an oil spill, on the other hand, is required to notify either the EPA or the Coast Guard immediately. The federal government then must act to clean up the discharge and to mitigate its effects. All costs of these operations are borne by the discharger. Dischargers may be fined for failing to give the required notice, for the discharge itself, and/or for violations of the equipment in operation in their facility.

thermal dischargers

Thermal dischargers are covered under a separate section of the FWPCA (Section 316). The EPA, in establishing standards for effluents other than heat, attempted to apply uniform standards. In the case of thermal dischargers, however, an exception is made in that the discharger is permitted to demonstrate that the effluent limitations proposed for the thermal component of the discharge are more stringent than necessary, to assure the protection of the various organisms present in the receiving water.

This exception is based on the assumption that although technology-based standards are desirable when applied to the majority of effluents, they are not applicable to thermal dischargers. This is due to the fact that heat dissipates, and that there are a small-enough number of thermal dischargers (power plants) to allow for a case-by-case review.

regulation of nonpoint sources

As noted previously, Section 208 or the FWPCA establishes a mechanism whereby state or regional agencies shall initiate regulatory programs to control nonpoint sources of pollution. According to federal guidelines, the governor of each state must designate areas with substantial water-quality-control problems. For each designated area the governor must name a representative organization to operate a continuing area-wide waste-treatment planning process.

The final plans must be submitted to the EPA for approval and must address the following nonpoint sources: (1) agricultural and silvacultural nonpoint sources, (2) mine-related nonpoint sources, (3) construction-related sources, and (4) saltwater intrusion resulting from the reduction of freshwater (surface or groundwater) flow. Once the management plan is approved by EPA, no permit may be issued in contradiction of the plan.

national goals

In Section 101 of the FWPCA, two national goals are stated: (1) that the discharge of pollutants into navigable waters be eliminated by 1985, and (2) that whenever attainable, an interim goal of water quality providing for the protection and propagation of fish, shellfish, and wildlife, as well as for recreation on and/or in the water, be achieved by July 1, 1983.

As might be expected, these provisions were among the most controversial of the 1972 Amendments. Consequently, this provision was not passed into law but was listed as one of the goals of the FWPCA. They are, therefore, not absolute requirements, since their application depends on technological and economic practicability.

marine protection, research and sanctuaries act of 1972 (ocean dumping act)

The Ocean Dumping Act was introduced and considered by Congress during the same period that the FWPCA 1972 amendments were under consideration. During this time many conflicts arose and, as a result, Congress passed both the Ocean Dumping and the 1972 FWPCA amendments with overlapping and conflicting provisions.

This act, from a water-management perspective, is not as all-inclusive as the FWPCA, since it attempts primarily to govern ocean dumping. The act regulated ocean dumping by requiring permits for various disposal operations. Dumping in all ocean waters lying seaward of the land and all internal waters (territorial or marine) are addressed by the Ocean Dumping Act.

Federal permits are required for the following operations: (1) the transportation from the United States of almost all materials for the purpose of disposal into ocean waters; (2) the transportation of these materials, from outside the United States, for disposal into the territorial seas of the United States; and (3) the transportation of these materials by a U.S. agency or official from outside the United States for the purpose of disposal into any ocean water.

safe drinking water act of 1974

In 1974 the Safe Drinking Water Act was passed by Congress. This act was an attempt to establish national drinking-water standards as well as to set limits on the concentrations of a variety of contaminants that may be present in water supplies. Table 10-1 is a partial list of the contaminants and their permissible concentrations in drinking water.

TABLE 10-1

federal drinking-water-contaminant levels

Contaminant	*Level (mg/liter)*
Inorganic chemicals	
Arsenic	0.05
Barium	1.0
Cadmium	0.010
Chromium	0.05
Lead	0.05
Mercury	0.002
Nitrate	10.0
Selenium	0.01
Silver	0.05
Organic chemicals	
Chlorinated hydrocarbons	
Endrin	0.0002
Lindane	0.004
Methoxychlor	0.1
Toxaphene	0.005
Chlorophenoxys	
2, 4-D	0.1
2, 4, 5-TP Silvex	0.01

*In addition, more than 20 halogenated and aromatic organics will be subjected to a special monitoring program to gather information for possible future standards.

CHAPTER 11

sampling apparatus

Fresh and marine water systems may be measured in situ (in place) or by collecting either the water or sediment and returning it to the laboratory for analysis. The type of information required will generally determine whether the system is to be measured in situ or whether the material is to be removed from the system for laboratory analysis. For example, if information on water velocity and direction is required, the water column must be measured in situ, whereas if various chemical parameters are to be determined, the water must generally be collected prior to analysis. In general, sampling and water measuring devices may be divided into five broad categories: water samplers, bottom samplers, biological samplers, current-measuring devices, and turbidity-measuring devices. Salinometers, oxygen meters, and selective ion meters are discussed in Chapter 12.

water samplers

In water sampling the problem of obtaining a water sample from a given depth and retrieving it without having the sample mix with water from other depths is encountered. In addition, if the water temperature is to be measured, there must be a means to accomplish this without having the temperature of the sample change as it is brought back to the ship. Specially designed *water bottles* are used to retrieve water from various depths. These bottles consist of open cylinders that are attached to a weighted wire (*hydrowire*) and lowered to the desired depth. The bottles are equipped with valves that are activated by a weight termed a *messinger*. The messinger is attached to the wire and released. It travels down the wire and hits a trigger on the water bottle. This causes the valves to close, thus collecting the water from that particular depth. After the messinger has tripped the valves (thereby collecting the water), the bottles are returned to the ship.

The most common water sampler in use is the *Nansen bottle*

(Fig. 11-1). The Nansen bottle is sent down to the required depth with the valves in the open position. At the desired depth a messinger is sent down, which strikes the trigger on the Nansen bottle. The trigger releases the top portion of the bottle from the wire and causes the bottle to reverse (turn upside down). As the bottle turns over (it is, however, held to the wire by the bottom clamp), the valves close, obtaining the water sample. Generally, several Nansen bottles are attached to the wire to allow for a series of samples to be taken at a variety of depths. As the first messinger strikes the topmost bottle, this bottle, as it reverses, releases another messinger, which travels down the wire and strikes the next Nansen bottle, causing it to reverse, take a water sample, and release another messinger, which then trips the third bottle; and so on. The process is repeated until the bottommost

FIGURE 11–1

Nansen bottles in rack. (*Photo: A. Haugen*)

Nansen bottle is tripped. This bottle has no messinger attached, since there is no Nansen bottle below to trip. In this manner, water samples at several different depths can be obtained almost simultaneously (Fig. 11-2).

The Nansen bottle is made of metal lined with Teflon; it generally holds 1 liter of water. Since the valves are metal, there is the possibility that corrosive salts will adhere to them. Unfortunately, this provides a means of contaminating successive water samples. To obtain samples for chemical analysis, bottles constructed of an inert plastic with few if any metal parts are preferred (Fig. 11-3).

The water samples collected by these bottles can be used for

FIGURE 11–2

Nansen bottle prior to taking a water sample. Note the position of the upper clamp. (*Photo: A. Haugen*)

FIGURE 11–3

Nansen bottle after taking a water sample. Note that the upper clamp has released and the bottle is in the reversed position. (*Photo: A. Haugen*)

various chemical and physical analyses. In addition, the particulate matter suspended in the sample can be removed by filtration and used in a variety of chemical, biological, or geological investigations. Since the volume of the water sample is accurately known, the concentrations of the various suspended and/or dissolved constituents comprising the sample can be accurately determined.

temperature measurements

Surface-water temperatures may be accurately measured by a simple thermometer. Measuring subsurface-water temperatures presents a problem, however, in that the temperature cannot be measured after the water is returned to the research vessel, because the temperature would change as the sample passes up through the various water levels, which can be expected to be at different water temperatures. Therefore, subsurface-water temperatures are generally measured either by reversing thermometers or with the bathythermograph.

reversing thermometers

Reversing thermometers are designed to obtain in situ subsurface-water temperatures. The thermometer (Fig. 11-4A and B) consists of a capillary tube drawn out into a loop (pigtail) and a constriction (appendix) immediately above the mercury reservoir. The thermometer is released in its upright position, and the mercury is able to pass freely from the reservoir into the capillary tube in response to the surrounding temperature. When the thermometer reverses (see below), the mercury column separates from the reservoir and is trapped in this position to give the temperature of the water at the instant of reversal.

The Nansen bottle is designed to hold reversing thermometers on their outer wall, and water temperature may be taken in situ at the same time that the actual water sample is obtained. Thus the temperature is taken by the thermometers at the same instant that the water sample is taken in the Nansen bottle. Generally, both a shielded and an unshielded thermometer are attached to the outer wall of the Nansen bottle and are sent to the desired depth. These thermometers are identical except that the shielded thermometer is completely enclosed in a protective glass case, is unaffected by pressure changes, and responds only to the temperature of the water at the depth to which it is sent. The unshielded thermometer is open at the bottom, allowing the surrounding water to enter the case and thus exert pressure on the thermometer. The pressure exerted on the unshielded

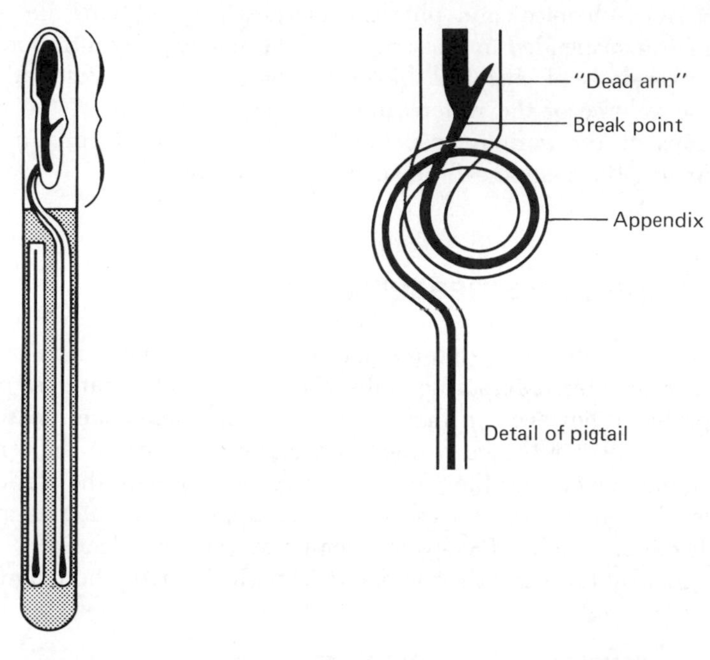

FIGURE 11–4

Shielded thermometer with detail of pigtail.

thermometer is, obviously, a function of the depth to which it is sent. As the Nansen bottle containing the thermometers reaches the desired depth, the thermometers are allowed to equilibriate prior to obtaining the water-sample and temperature information. In the shielded thermometer the mercury responds only to the temperature of the surrounding water, whereas in the unshielded thermometer the mercury responds to both the water temperature and the hydrostatic pressure.

As the messinger trips the Nansen bottle it reverses, collecting the water sample. As noted above, when the thermometers reverse, the mercury column separates from the mercury contained in the reservoir and falls to the opposite end of the thermometer. Since it is detached from the reservoir at this stage, it no longer is able to respond to temperature changes. The shielded thermometer gives the actual in situ temperature, whereas the unshielded thermometer gives a temperature reading as affected by the pressure.

It is known that the pressure exerted by subsurface water is a function of depth and will cause the unshielded thermometer to give a false (high) temperature reading. Therefore, a comparison of the actual temperature (from the shielded thermometer) with the false temperature combined with the density of the surrounding water (obtained from the collected water sample) enables the depth at which the water sample is collected to be calculated. This is a definite advantage, for, although the depth may be calculated by measuring the amount of wire (attached to the Nansen bottles) that has been played out, this gives only an estimate of the actual depth. This is due to the fact that currents, vessel drift, and wind will cause the wire to descend on an angle, thus giving a false measurement of the actual depth at which the sample was taken.

bathythermographs

The *bathythermograph* is illustrated in Figure 11-5. Although this instrument is not as accurate as the reversing thermometer, it

FIGURE 11–5

The bathythermograph. (*Photo: A. Haugen*)

allows continuous measurement of temperature versus depth and permits these data to be collected more rapidly.

The bathythermograph consists of a temperature-sensitive element that scribes a tracing on a coated-glass slide in response to changing temperatures. The slide itself is mounted on a pressure-sensitive spring-loaded bellows that contracts in response to changing temperatures over a slide that moves in response to changing pressures (due to and correlated with water depths). These two motions cause the stylus to trace a temperature–depth curve on the slide as the bathythermograph is lowered into the water column. Upon retrieval of the instrument, the slide is removed, placed against a calibrated grid, and "read." The slide may be retained and thus provide a permanent record of temperature versus depth at a given station.

bottom-sampling devices

There are three basic devices utilized to obtain bottom samples: dredges, grabs, and coring devices. Both dredges and grabs are able to take samples for both geological and biological analysis but are incapable of obtaining undisturbed samples. Coring devices, on the other hand, are able to obtain undisturbed bottom samples that allow the investigator to study sediments in the order in which they were initially deposited. Corers do not, however, obtain samples of sufficient quantity to permit biological analysis of organisms other than the microfauna (*Foraminifera*).

dredges and grab samplers

A *dredge* is a boxlike apparatus that is lowered onto the bottom and dragged behind a ship. As it moves over the bottom, the dredge scrapes biological samples from the surface and digs into the softer sediments, obtaining, by this means, a sample of bottom organisms, small rock, sand, and mud. A wire or cloth mesh liner is placed inside the dredge to permit the free flow of water but to retain the sediment and biological materials that enter. The mesh size determines the size of the material that is collected. Since a dredge tends to skip and bounce over the sediment as it is towed, it does not obtain a quantitative sample, and, as noted above, the sample tends to become well mixed in the process.

Grab samplers are available in several forms. The basic design, however, consists of a pair of jaws and an upper reservoir to hold the sample. Grabs are generally released from the ship with the jaws open;

as the sampler strikes the bottom, the jaws are "tripped" and snap shut, obtaining the sample. Water that washes through the reservoir as the grab is returned to the surface tends to stir and disturb the collected sediment. In addition, pebbles or shells often tend to keep the jaws partially open so that the sample is washed out and lost during retrieval.

coring devices

Coring devices, like grab samplers, are available in several different designs and modifications. In its simplest form (Fig. 11-6), a coring device consists of a hollow metal tube with a weighted upper end. The tube is designed to separate toward the middle to allow the insertion of a plastic core-tube liner into its lower section. Attached to the bottom of the metal tube is a sharp cutter head, which facilitates the passage of the tube into the sediment. Attached to the upper end of the metal tube is a flap that serves to create suction on the system after the sample is obtained. The entire apparatus is attached to a line or cable aboard a ship, and the flap is placed in the open position. The coring device is then released and allowed to freefall through the

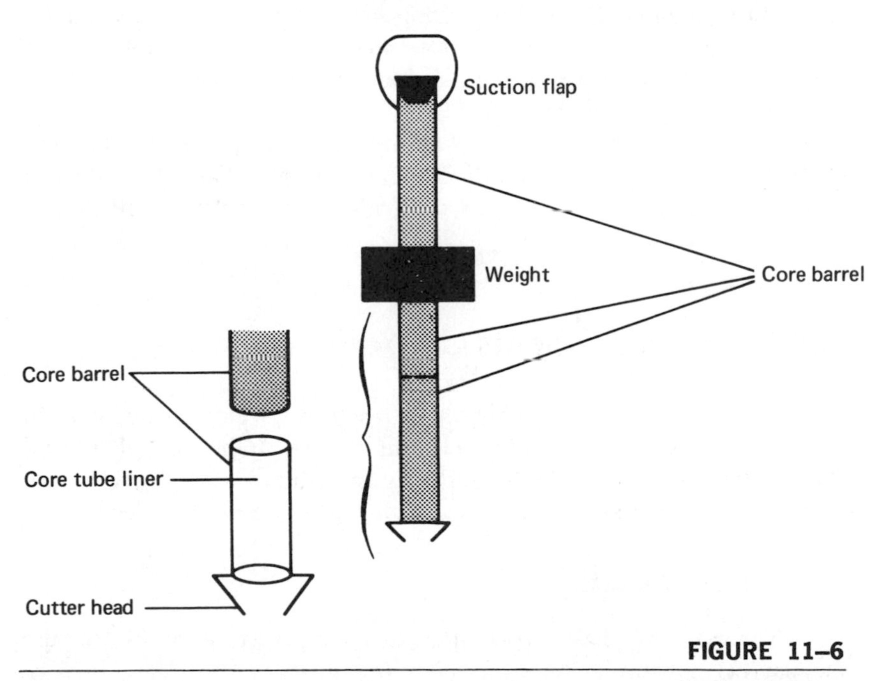

FIGURE 11–6
Coring device.

water column. As it strikes the sediment, the weight drives the core barrel into the substrate, and the sediment enters the core-tube liner. Simultaneously, the flap snaps shut and creates suction on the system. The suction allows the sediment to be held firmly in the core-tube liner. The apparatus is then returned to the ship and the liner holding the sediment is removed from the barrel and saved for analysis.

biological samplers

Bottom organisms are generally obtained by dredging and grab sampling. Organisms that inhabit the water column (plankton, fish, etc.) are generally collected in nets of various meshes. The mesh size, obviously, determines the size of the organism collected. Some nets are designed to be sent down to a given depth in the closed position. At the required depth a messinger is sent down to open the net, which then collects the desired sample. After the sample is collected, the net may be reclosed and brought back to the ship in this position. In this manner only organisms that inhabit a given depth are collected. Many nets (especially plankton nets) contain a propeller-device meter attached to the front of the net. The water passing into the net turns the propeller on the meter and gives an accurate measure of the water passing through the net. In this way the number of organisms per unit volume of water can be calculated.

A net termed a *bottom trawl* is often used in lieu of dredges and grabs to obtain large bottom-dwelling organisms. The bottom trawl consists of large mesh nets that drag along the bottom and collect whatever is in their path.

current measurements

There are several means available to measure water movement. In marine systems and large lakes, current meters, fluorescing dyes, and float bottles are generally employed. In groundwater systems the movement of fluorescing dyes or chloride ions is generally traced.

current meters

There are two basic types of current meters presently in use: the mechanical current meter and the more modern electronic meters. The *mechanical current meter* consists of a current rotor and a system of dials. The rotor spins in response to the current and releases a bronze pellet after completing every 30 revolutions. The pellets fall into compartmentalized slots which correspond to compass points and are used to indicate direction. Thus the number of pellets released indicates the total number of revolutions completed by the propeller (water speed), and the distribution of the pellets is used to indicate water direction.

Each series of measurements using a mechanical meter of this type entails removing the meter from the water, recording the data (the number of pellets released and the distribution in directional compartments), resetting the meter, and placing it at the desired depth for the next series of measurements. In addition, no information on the variability of speed can be obtained.

Electronic current meters generally consist of a rotorlike device that is turned by the passing water. As the rotor turns, it generates an electric current which is transmitted to the ship and is used to drive a meter or chart recorder pen. A magnetic or gyroscopic compass is also attached to either the current-meter body or to the directional vane. The orientation of the compass is electrically sensed and is also transmitted to the ship, where it is presented as a meter reading or as a strip-chart recording. Electronic current meters give a continuous measurement of current velocity and direction.

drift bottles

Drift bottles give the average direction and speed of the currents in which they are placed. A drift bottle is a sealed container that is weighted to be of the same density as the water mass to be traced. The weighted drift bottle is placed in the appropriate water mass and released. Inside each drift bottle is a card that requests the finder to fill in the date and place of recovery and return the card to the institution performing the study. When the card is returned, the average speed and direction taken by the bottle can be determined. Plastic cards, serving the same purpose, are often used instead of drift bottles at present.

dye studies

Fluorescing dyes are often released into water masses and used to trace the path of these waters. These dyes are readily visible in high concentrations and are easily detected by using a fluorometer in trace amounts (ppm and ppb). Generally, the dye is mixed with water and prepared to be of the same density as the water mass in which it is to be released. After the dye has been released, a ship will set up a series of stations from the point of dye input, and water samples will be obtained and analyzed fluorometrically (see Chapter 12).

use of chloride ions

In many groundwater systems it is found that dissolved ions, such as iron, fluoresce over a very wide range and interfere with the detection of the fluorescent dye customarily used. In these cases sodium chloride is used in place of the dye. The sodium chloride, when dissolved in water, dissociates into its component ions (Na^+ and Cl^-), and these ions will tend to travel as a part of the water mass in which they are dissolved. Since normal groundwater contains no chloride ions, any chloride detected in test wells downgradient of the point of input can be assumed to have been placed there as a part of the study. The presence of chloride ions is easily determined by silver nitrate titration (see Appendix III).

turbidity-measuring devices

The degree of water transparency (*turbidity*) is of importance since it indicates the amount of suspended solids in the water column and indicates the extent of the euphotic zone. The turbidity of water is commonly measured manually by a *Secchi disk* (Figs. 11-7 and 11-8) or electronically by a hydrophotometer.

The Secchi disk is a white, or black-and-white, disk that is lowered into the water column on a line marked off in meters. The disk is lowered until it disappears from view and is then returned to the ship. As it is returned, the number of meters of line that had initially been played out is determined and noted. This indicates the transparency of the water column.

Transparency may also be determined by using a hydrophotometer. This device consists of a light source and a photo cell and is similar, although simpler, than a spectrophotometer (Chapter 11). Light passes through the water sample and strikes the phototube. The phototube converts the light energy to electrical energy, which is then used to drive a meter calibrated in percent light transmission. As the amount of dissolved material increases, less light strikes the phototube and the light transmission of the sample is decreased.

FIGURE 11–7

The Secchi disk. (*Photo: A. Haugen*)

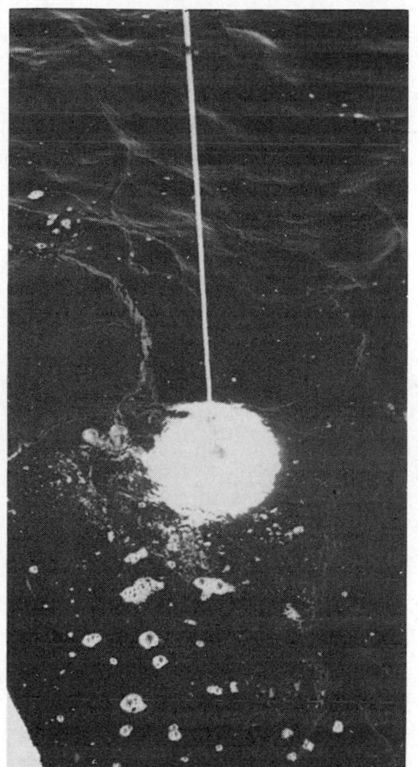

FIGURE 11–8

The Secchi disk beneath the water surface. (*Photo: A. Haugen*)

SUGGESTED READINGS

Barnes, H. 1959. *Oceanography and Marine Biology: A Book of Techniques.* London: Macmillan & Company Ltd.

Issacs, J. D., and C. O. Iselin. 1952. *Symposium on Oceanographic Instrumentation.* Washington, D.C.: National Academy of Sciences–National Research Council, Publication 309.

U.S. Naval Oceanographic Office. 1968. *Instruction Manual for Oceanographic Observations.* Washington, D.C.: Government Printing Office.

QUESTIONS

1 / Explain the value of obtaining readings on both shielded and unshielded thermometers when measuring temperatures in deep water.

2 / Explain the operation of the bathythermograph.

3 / Explain the operation of the gravity corer.

4 / Discuss the differences in the operation of the two basic types of current meters.

5 / Discuss the various methods of tracing water currents.

CHAPTER 12

analytical instrumentation

Seldom, if ever, is an analyst expected to determine the entire spectrum of materials that may be present in a water sample. Rather, the sample is analyzed to determine only the presence or absence of a specific component or components, and the analyst always has a specific, well-defined plan in mind when the analysis is initiated. It is necessary to proceed with an analysis from this standpoint since the sample size is generally limited, specific sample pretreatment may be necessary, and specific instruments may be required to perform the various types of analysis.

Two basic types of chemical analysis may be carried out on water samples: qualitative analysis and quantitative analysis. In *qualitative analysis* only the presence or absence of a specific substance is of interest, and the exact determination of the concentration of that substance is not required. In *quantitative analysis,* on the other hand, the exact concentration of a particular material is required. Generally in water-quality work, quantitative analysis is routinely performed, and qualitative studies are a "by-product" of the quantitative step. For example, if a series of stations in a given estuary are to be analyzed for copper, the analysis will be of a quantitative nature. If any of the samples do not show the presence of copper, the results will be reported as zero or "below-detection limits," depending upon the sensitivity of the analytical method employed.

As noted previously, it is absolutely necessary to have a well-defined idea of the materials to be determined, since this will determine the sample size, the actual course of the analysis, and the instruments that are necessary. The most common instruments used in water analysis are spectrophotometers, fluorometers, gas chromatographs, and selective ion meters, including pH and oxygen meters. This section deals only with the basic theory involved in the above-mentioned analytical instruments. Specific methods of analysis are omitted and are discussed in the appendixes.

spectrophotometry

general principles

Various components dissolved in a given water sample are commonly analyzed by passing different types of light (visible, infrared, or ultraviolet) through the water sample and measuring the effects of the dissolved components on the light. The common instruments used in this type of analysis are *visible spectrophotometers* and *atomic-absorption spectrophotometers.* These spectrophotometers measure the amount of light absorbed by the component under investigation. In both methods the mere absorption of light by the components under consideration will give a qualitative analysis (will indicate the presence of the material under investigation). In each method the actual amount of light absorbed will give a quantitative analysis, since the amount of light absorbed is dependent on the actual concentration of the dissolved material in solution.

Both ultraviolet and infrared spectrophotometers are also used in water analysis, although not as extensively as the visible and atomic absorption spectrophotometers. These instruments, as their names imply, use different types of light but are basically similar in the instrumentation and mode of operation.

properties of light

All light, more properly termed electromagnetic radiation, travels from its source in a wave motion and is, therefore, described in terms of wavelength or frequency. *Wavelength* is defined as the number of waves passing a given point in a given period of time. Different types of light will move at different velocities and will have different wavelengths. Thus the various wavelengths are used to describe the type of electromagnetic radiation (Fig. 12-1) emitted from a given source. In the case of visible light, the various wavelengths are equivalent to the various visible colors, whereas in the ultraviolet and infrared regions of the spectrum, the wavelength can be considered to indicate the energy of that particular type of light. The units used to indicate the various wavelengths of light are micrometers and nanometers. A micrometer is equivalent to 10^{-4} centimeter (cm) and a nanometer is equivalent to 10^{-9} cm.

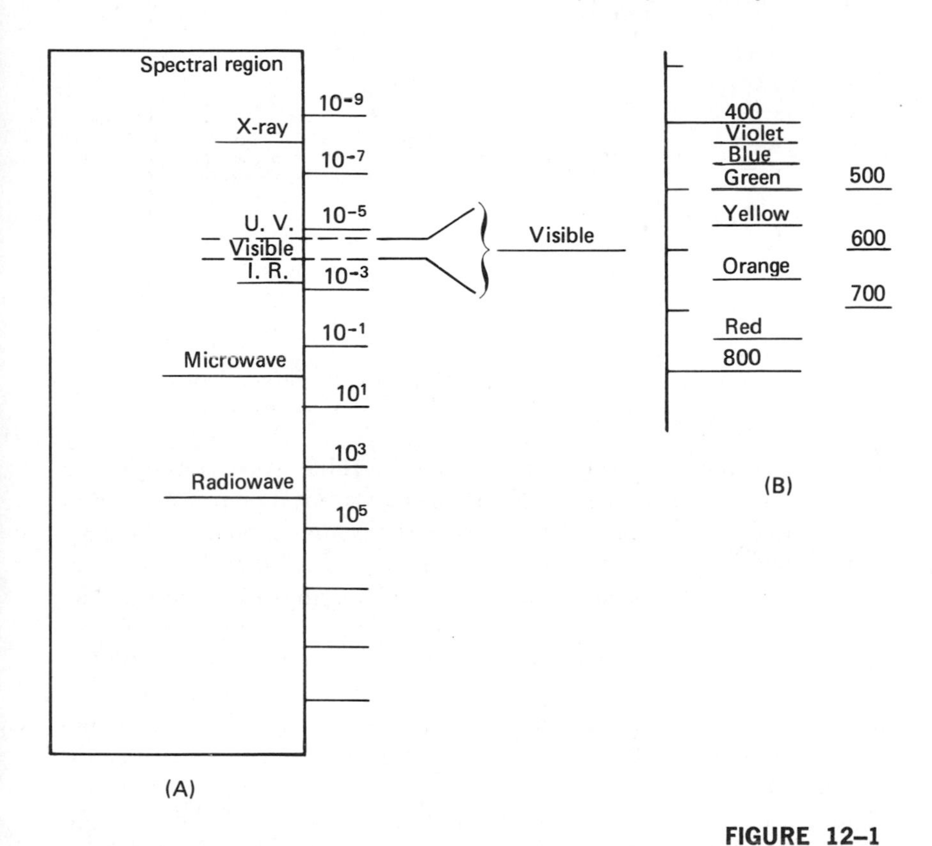

FIGURE 12–1

(A) The electromagnetic spectrum (wavelength in cm). **(B)** Expanded visible region (wavelength in millimicrons).

absorption of light

Basically, the absorption processes described below for the visible region of the electromagnetic spectrum also occur in the ultraviolet and infrared regions. Since the light in these regions is not visible, the results, although similar, are not as easily visualized. Therefore, although the following discussion is confined to the visible spectrum, it is to be noted that the same basic principles of absorption occur in the nonvisible regions.

White light is considered to consist of all the visible colors and is termed *polychromatic light* (many-colored light). When polychromatic light is passed into a colored solution, it is found that a portion of the light is absorbed by the components comprising the solution and a portion passes through with no interference. Further examination will

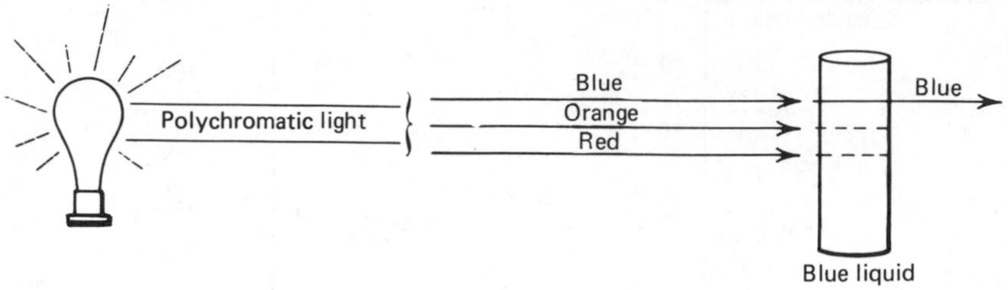

Polychromatic light
Blue
Orange
Red
Blue
Blue liquid

FIGURE 12–2

The conversion of polychromatic light to monochromatic light.

reveal that only a portion of the total visible spectrum (a small number of wavelengths) is absorbed to its maximum extent. For example, if polychromatic light is passed through a blue liquid, blue light would be observed exiting from the liquid (Fig. 12-2). This indicates that the blue solution absorbs wavelengths (light) other than blue to a greater or lesser extent and allows only the blue light to pass through totally unhindered. If, however, the polychromatic light is passed through a red filter prior to entering the blue solution (Fig. 12-3), the red filter would convert the polychromatic light to monochromatic light (light of one color) by absorbing all the wavelengths but red, which would be passed through the filter and into the blue liquid. The result would be the complete absorption of the red light by the blue liquid, since red light is maximally absorbed by the blue. This is similar to holding a red filter up to a polychromatic light source and observing the resulting red light pass through the filter, and then repeating the operation with a blue filter. If the operation is performed

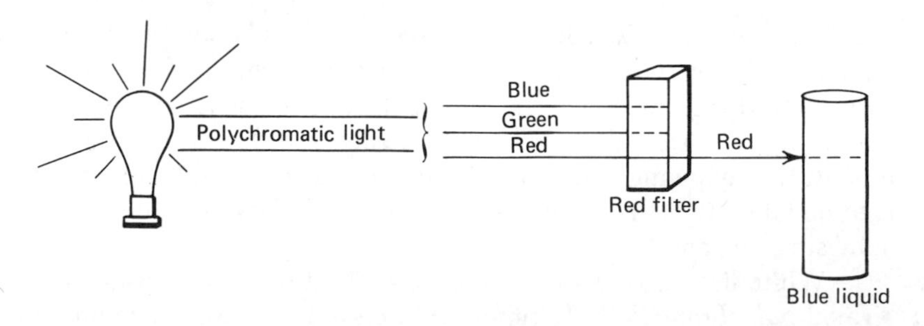

FIGURE 12–3

The conversion of polychromatic light to monochromatic light.

a third time by superimposing the red and blue filters, no light would be observed to pass through the filter. This indicates that the red wavelengths are absorbed maximally by the blue, and vice versa.

visible spectrophotometry

In *visible spectrophotometry* the absorbance of light by the dissolved materials under investigation is primarily utilized in quantitative analysis. The instrument, schematically represented in Fig. 12-4, consists of a polychromatic light source, a mirror to direct the light into the instrument, a monochrometer or filter to convert the polychromatic light to monochromatic light, a slit to direct a portion of the light into the sample compartment, and a phototube that picks up the light as it passes from the sample and converts the light to electrical energy. The electrical energy produced by the phototube is then used to drive a meter to indicate the amount of light that was passed through the sample. Since the light entering the sample is considered to be the maximum amount (100% transmitted and entering the sample), the difference between the light entering the sample and that exiting is the amount that was absorbed by the components dissolved in the sample. The meter can be used to give readings of either percent of light transmitted (% transmission) by the sample or the amount of light absorbed (absorbance) by the sample. In other words, a sample with a high % transmittance would have a low absorbance reading, whereas a sample with a low % transmittance would have a high absorbance of light.

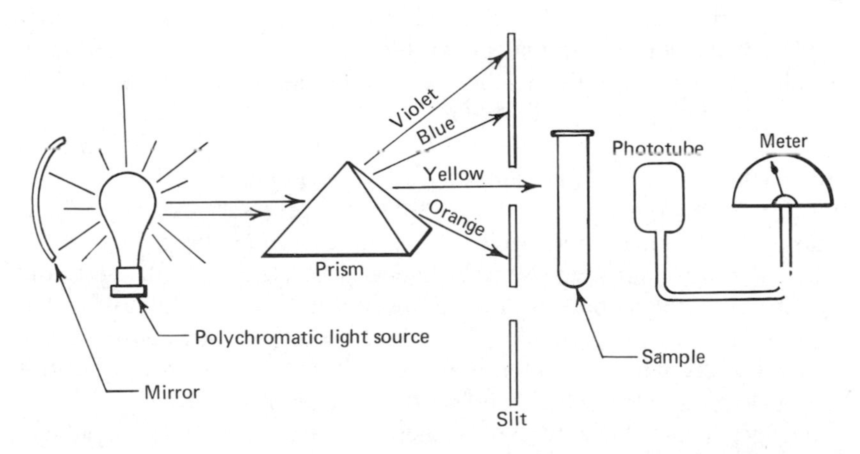

FIGURE 12–4

Schematic diagram of a visible spectrophotometer.

The *monochrometer* is a device that serves to refract the light and split it into its component colors (wavelengths). Generally, the setting of the monochrometer can be changed by a dial (calibrated in the wavelength units of micrometers or nanometers) on the face of the instrument. Some instruments use filters or diffraction gratings. The sample is placed in a specially designed test tube, termed a *cuvette,* the walls of which are of uniform thickness to prevent the indiscriminant scattering of light. The more sophisticated spectrophotometers use specially made optically matched cuvettes for extreme accuracy. Cuvettes of various capacities are available, depending upon the specific analysis required.

Analytical methods / The principles underlying spectrophotometric methods of analysis are based on two facts: (1) that various substances will absorb light maximally at specific but different wavelengths, and (2) that as the concentration of a particular substance increases, the absorbance of light will increase accordingly; conversely, the amount of light passed through (% transmittance) will decrease.

Prior to performing quantitative analysis, it is necessary to determine the wavelength that is absorbed maximally by the material that is suspected to be in the sample and, thus, under analysis. For example, if a sample or series of samples are to be analyzed for the presence of cobalt (generally present in its ionic form), a known solution of cobalt would first be prepared. This is usually accomplished by dissolving a cobalt salt such as cobalt chloride in distilled water, placing this solution in a cuvette, into the sample compartment, and exposing it to all the various wavelengths of visible light. Since cobalt ions give the resultant solution a reddish color, it would be found that in the wavelengths equivalent to blue light a greater absorbance would be obtained, while in the wavelengths equivalent to red light a lesser absorbance would be obtained. Similarly, if a sample is to be analyzed for copper, which imparts a bluish color to the solution, a known solution of copper sulfate would be prepared and exposed to all the visible wavelengths. In this case the reverse would be noted: maximal absorbance in the red regions and minimal absorbance in the blue. If wavelength versus absorbance were graphed for both the copper and the cobalt solutions, very different graphs (Fig. 12-5) would be obtained. These graphs, termed *absorbance spectra,* are generally prepared to determine the proper wavelength to utilize when carrying out a specific analysis. After the proper wavelength has been determined, the monochrometer is set accordingly and the quantitative analysis can begin.

Quantitative analysis involves the measurement of the absor-

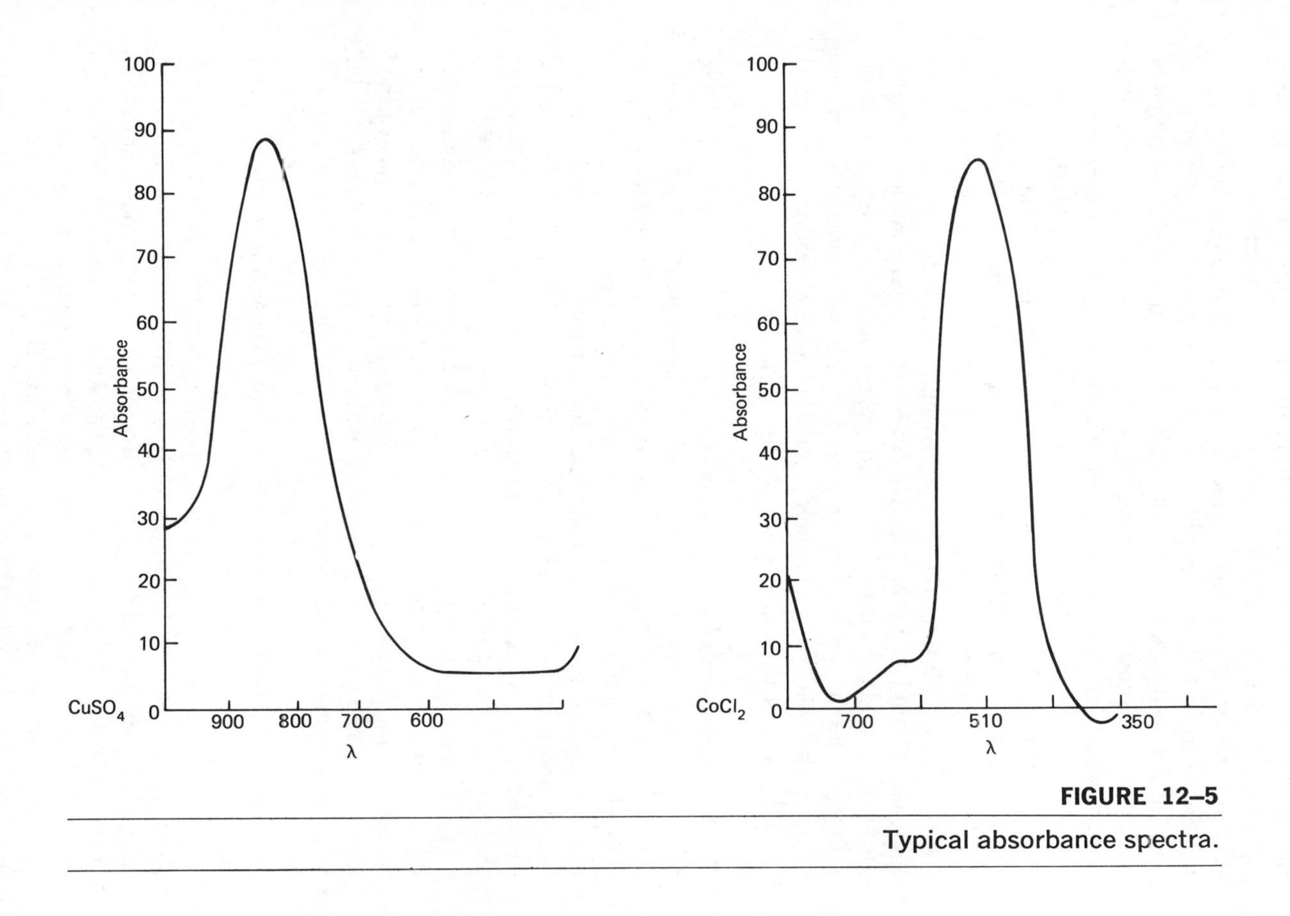

FIGURE 12–5

Typical absorbance spectra.

bance of a specific wavelength of light by the material under analysis. As noted previously, the concentration of the various compounds in solution will determine the degree to which this monochromatic light will be absorbed. For example, a solution containing 1 mg/ml of a specific material will absorb only half of the amount of light that a 2 mg/ml solution will absorb, and so on. Based upon these facts, it is necessary to standardize the spectrophotometer prior to carrying out the analysis. Standardization merely involves the preparation of a series of standard known solutions of different concentrations. After the standard solutions have been prepared, the instrument is "blanked" (p. 185), each solution is run through the instrument at the predetermined wavelength, and the amount of light that is absorbed by each standard is noted and graphed. This graph is termed a *standard curve.* Once the standard curve (for each particular substance) has been prepared, the unknowns may be placed in the instrument, the amount of light absorbed by each noted, and the concentrations determined from the standard curve.

Practical considerations / In visible spectrophotometry most of the elements, many ions, and innumerable compounds dissolved in water can be quantitatively analyzed. Although some of these materials are self-colored and can be analyzed directly with no preparatory steps, it is generally necessary to develop the desired color by the addition of one or more color-forming reagents, termed *chromogenic reagents.* The chemical processes used to prepare suitable colored solutions for analysis are termed *chromogenic reactions.*

The chromogenic process is chemical in nature and requires careful attention, since the results will not be valid unless the colored solution that is measured has been prepared properly. The chemical problem is to transform all the desired constituents into their colored state. With an extremely small quantity of material, this conversion may not be quantitative or rapid, and usually a large excess of chromogenic reagent is required to ensure a complete conversion to the colored state.

Ideally, the chromogenic reagent should itself be colorless so that an excess of reagent does not interfere with the light absorbance by the material under analysis. If the chromogenic reagent possesses some degree of color, the amount of reagent added must be scrupulously controlled. In addition, the effect of other substances that may be present in the sample must be known, since they may (1) interfere by developing a color with the chromogenic reagent, (2) precipitate and physically prevent the light from passing through the sample, and/or (3) inhibit the chromogenic reagent by consuming it and thus decreasing the amount of chromogenic reagent available to react with the desired constituent.

Colored solutions suitable for use in visible spectrophotometry should have the following properties: (1) the solution should be intensely colored so that small changes in concentration cause an easily detectable change in intensity and thus in the absorbance of light; (2) both the color and intensity of the resultant solution must be reproducible so that the analysis may be rechecked if necessary; and (3) the intensity of the colored solution should remain constant long enough to enable the measurement to be made.

After the chromogenic reaction has been completed, the samples are ready for analysis. It is to be noted, however, that both the excess chromogenic reagent and the solvent (sea, fresh, or groundwater) may absorb light at the wavelength used in the analysis. Any absorbance due to the chromogenic reagent is overcome by *blanking* the instrument. Blanking involves the preparation of a cuvette with the required volume of distilled water, replacing the sample, and adding the proper amount of chromogenic reagent. The cuvette, then, contains chemically pure distilled water and chromogenic reagent. There should be no development of color since the constituent under analysis in the actual samples is absent in the distilled water. The cuvette is placed in the sample holder, and the instrument is physically adjusted to give a zero-absorbance reading as the light is passed through the distilled water–chromogenic reagent solution.

In the case of a natural sample (sea, fresh, or groundwater), the samples may be both highly colored (due to the presence of constituents other than those under consideration) and contain suspended material. Both the color and the suspended material may interfere with the light that must be passed through the sample in the course of the analysis. To remove the particulate matter, the samples are routinely filtered upon collection. To account and compensate for the natural "background" color of the sample, the instrument is initially blanked, as noted above. After blanking the water to be analyzed is placed in the instrument *with no chromogenic reagent added,* and the absorbance at this point is noted. When this has been accomplished, the chromogenic reagent is added to the sample and the color is allowed to develop. After the color development is complete, the sample is placed in the instrument and the absorbance noted. At this point the absorbance due to the background color can be subtracted from the absorbance of the sample plus the chromogenic reagent. This resultant absorbance is used to determine the actual concentration from the standard curve. Instrument operation is discussed further in Appendix II.

atomic-absorption spectrophotometry

Atomic absorption is a method used to analyze specific elements that may be present in a particular sample. Unlike spectrophotometry, this method is capable of measuring only the presence of elements in their free state and cannot distinguish between, or measure the presence of, specific ions, molecules, and the like. For example, in visible spectrophotometry it is possible to determine the concentration of both inorganic phosphate (PO_4) and organic phosphate in water samples. In atomic absorption only the presence of specific elements can be determined. The element under investigation is released from the surrounding solution or the bonds holding that particular element in its molecular or ionic configuration. In its free state the element is capable of absorbing specific wavelengths of radiation emitted from a light source within the instrument. The remainder of the light passes through a monochrometer, which allows only the specific wavelength known to be absorbed by the element to pass on to the phototube, where it is converted to electrical energy. This energy is then used to give a meter reading. Thus this method is similar in theory to visible spectrophotometry. The instrument is illustrated schematically in Fig. 12-6.

Practical considerations / Atomic-absorption spectrophotometers are

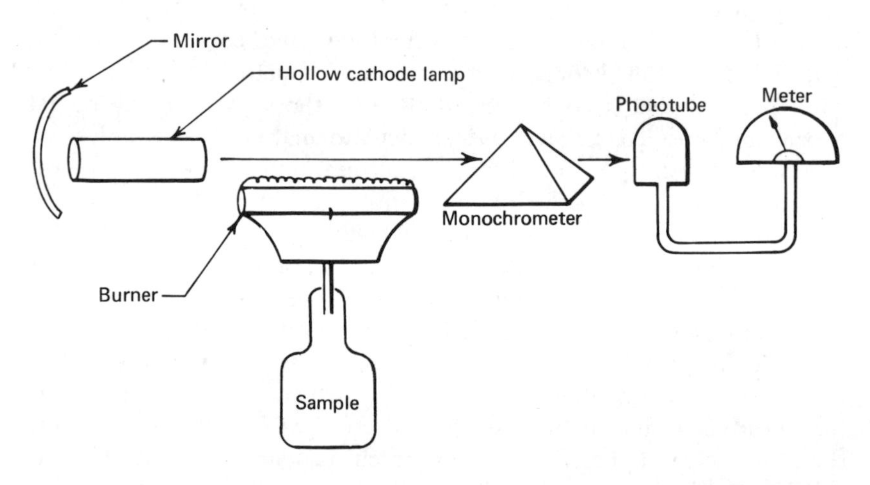

FIGURE 12–6

Schematic diagram of an atomic-absorption spectrophotometer.

blanked and standardized in a manner similar to that described for spectrophotometric analysis. Since, however, it is the absorption of light by free atoms that is measured, there is no need to develop color by the use of chromogenic reagents. All that is necessary in the standardization is the preparation of an accurate set of standard solutions. In addition, owing to the nature of the analysis, very low concentrations of elements can be detected by this method. It is important to note that, as above mentioned, only elements can be determined by this method.

Instrumentation / In atomic absorption the elements under investigation are placed in their free (vapor) state by burning the total sample in a flame. This is accomplished by aspirating the sample, which is carried into the flame and vaporized there. The sample then passes into the light path and portions (specific wavelengths) of this light are absorbed by the particular atoms under study. The light source in atomic absorption is provided by a lamp filled with neon or argon gas at low pressures and a cathode which contains the element that is being determined in the sample. This light source is termed a *hollow cathode lamp*. When current passes through the lamp it causes the atoms (contained within the lamp) to move from the cathode into the path of gas, where collisions between the gas and these atoms cause the atoms to gain energy and to be raised to a higher energy state. In this state they become unstable and lose the excess energy gained in the collisions by emitting this energy as light. The light that is emitted is the light that passes through and is absorbed by the vaporized sample. Since this light was emitted by atoms identical to those under investigation, this light will be readily absorbed should any of these specific elements be present in the vaporized sample traveling into the light path. For example, if seawater is to be analyzed for the presence of lead, a lead lamp would be placed in the instrument to serve as the light source. The seawater would then be aspirated, pass into the flame, and the entire sample (the seawater and all of the dissolved elements) would be placed in the vapor state. During this operation the light source simultaneously emits radiation from the excited lead atoms in the cathode tube (lead lamp). This radiation will pass through the sample and, since the light was emitted by only lead atoms at a rather specific wavelength, only this light is capable of being absorbed by the lead atoms (if any) in the sample. The unabsorbed light passes through a monochrometer and ultimately strikes a detector, giving a meter reading that can be converted to concentration by using the previously prepared standard curve. Many of the newer instruments read out directly in concentration. Atomic absorption methods are widely used in the analysis of trace metals in both water and sediment samples. Specific methods of analysis are given in Appendix II.

fluorometric analysis

In fluorometry water samples are exposed to ultraviolet light, which is absorbed by the material under study and serves to raise these materials to a higher energy level. The material then returns to its normal energy state by losing this energy in the form of light which is emitted at either the same or a longer wavelength. The emitted light then passes into the optics of the instrument, to ultimately strike a detector and give a meter reading which can be converted to concentration (generally by use of a prepared standard curve).

general principles

In this method the amount of light emitted by a specific material is measured. Consequently, if the material under analysis is absent from the sample, no light will be emitted, and zero concentration will correspond to absolute darkness. Thus in fluorometry sensitivity depends upon the detection of the light emitted by the dissolved materials that have previously been excited. This is in contrast to spectrophotometry where the amount of total light absorbed by a material is measured. In spectrophotometry the sensitivity in the low concentration ranges depends on the detection of the small fractional decrease in maximum light concentration. Fluorometric methods are capable of routinely measuring in the parts per million and parts per billion ranges of concentration. In one instance, measuring the movement of ocean currents, it was possible to detect a fluorescent dye in the extremely low concentration of 1 part dye to each 10^{12} parts seawater.

instrumentation

The instrumentation (Fig. 12-7) involved in fluorometric analysis consists of a light source which emits a wide spectrum of ultraviolet light. A primary filter allows only the proper wavelengths of light to enter the sample compartment. This light enters the sample and brings about the necessary excitations of the constituents within the sample. As noted previously, the sample spontaneously loses this excess energy as light, which then passes through another filter, termed the secondary filter. This filter blocks extraneous light and passes only the light

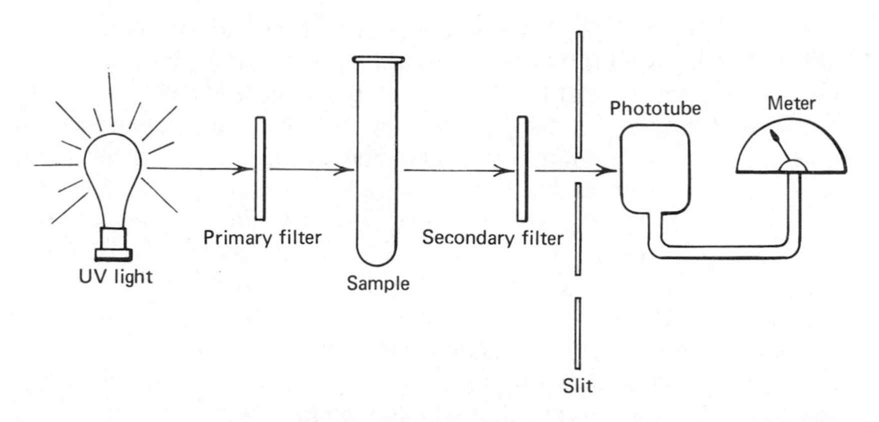

FIGURE 12–7

Schematic diagram of a filter fluorometer.

emitted by the material under study. Thus the function of the secondary filter is to eliminate nonessential light and allow only those wavelengths emitted by the material of interest to strike the detector. The detector converts this to electrical energy which is used to give a meter read-out.

practical considerations

This method depends upon the ability of the material under study to fluoresce (emit light when excited). Although few materials fluoresce strongly, the vast majority of materials present in natural systems do exhibit some degree of fluorescence. This is both an advantage and a disadvantage, since in many cases fluorescence due to undesired materials will have to be eliminated. Because there is a choice of not only the wavelengths that enter the sample (via the primary filter) but also of the wavelengths that are permitted to strike the detector (determined by the secondary filter), there are two means whereby it is possible to avoid or eliminate indiscriminant fluorescence. For example, if a fluorescent dye is used in the course of a current study, it may be found that other constituents in the water sample will excite at the same wavelength as the dye to be analyzed. If this is the case, light passing through the primary filter will raise not only the dye but the other components to a higher energy level. These materials, as well as the dye, will lose this excess energy as light.

In most cases, however, the light emitted by the dye will be at a different wavelength than the light emitted by the other components. If this is indeed the case, the proper secondary filter can be selected and this filter will pass only the light emitted by the dye and absorb the light emitted by the other fluorescent materials. Thus only the desired light will strike the detector and give a meter reading. In other cases it may be found that both the dye and other materials in a sample will emit at the same wavelength but excite at different wavelengths. In these cases the proper primary filter can be selected to pass only light capable of causing fluorescence by the dye into the sample. Since only the dye is emitting light, only that light is striking the detector. There are, however, many cases where other materials fluoresce at such large numbers of wavlengths that it is impossible to segregate the light, and in these cases this method cannot be used. For example, groundwater on Long Island contains many iron compounds that both excite and fluoresce over extremely wide wavelengths. In these cases it is impossible to segregate the light emitted by the dye from natural background fluorescence; therefore, in this region dye studies of this type cannot be used. In some cases, however, it may be possible to compensate for background fluorescence by the method described on page 185.

As noted previously, fluorometry, under the proper conditions, is a valuable technique in tracing the movements of water masses such as currents. In addition, it can be used in tracing the pathways of various contaminants from their source of input into water bodies.

chromatographic methods

Chromatography is a method used to separate a mixture or solution into its individual component parts. Although the major chromatographic method used in water-management studies is gas chromatography, the original, less complex, chromatographic procedure, termed *column chromatography,* will be discussed first, since much of the theory and terminology can be applied to gas chromatography.

column chromatography

The original chromatography procedure involved the separation of leaf pigments. As a preparatory step the leaves were ground in a suitable solvent such as petroleum ether, and the resultant solution was passed down a glass tube packed with powdered chalk. As the pigments entered the column it was observed that, initially, they were retained on the chalk at the top of the column. As additional solvent (petroleum ether with no dissolved pigments) was added and passed down into the column, the green pigments began to move down into the column along with the solvent and separate into their individual, component parts. Eventually the component pigments separated to such an extent that the column of chalk (with the separated pigments) could be removed from the column. The different pigment zones could be cut out, dissolved from the chalk, and pure samples of each pigment could be obtained. These fractions could then be subjected to additional analysis if necessary. For example, in many cases a solution is separated into its component parts by chromatographic methods and then each component is analyzed quantitatively by spectrophotometric methods.

Based upon this method, therefore, chromatography is presently defined as a method for separating the components of a mixture or solution on a stationary support (in the example above, the chalk). The separation is accomplished by means of a mobile phase (the petroleum ether in the example above) that carries the components at different rates along the support. Separation is accomplished due to the following:

1 / The various components are soluble to a greater or lesser extent in both the stationary phase and the solvent (mobile phase). Thus a component that is very soluble in the stationary phase and only slightly soluble in the solvent will move down the column at a slower rate than a component only slightly soluble in the stationary phase and very soluble in the mobile phase.

2 / The physical size of the stationary phase and the length of the column affects the efficiency of separation. It has been demonstrated that columns packed with identical stationary phases of different sizes will show different separation rates. For example, if one column is packed with finely powdered chalk and the other packed with coarse, granular chalk, the stationary phase of the smaller particle size will give more efficient separation. In addition, extending the length of the column will also give better separation since this merely increases the amount of time that the material to be separated remains in contact with the stationary phase. It is to be noted, however, that decreasing the particle size and/or increasing column length decreases the flow rate of the solvent/sample and thereby increases the time required to complete the analysis. In the case of an analysis requiring the separation of closely related materials, this may be the only means to accomplish a separation, and the speed of the analysis may have to be sacrificed to effect an efficient separation.

At present, rather than extruding the column packing and dissolving the separated components as discussed above, the mixture is allowed to separate completely in and pass through the column. As the material leaves the column, the individual components are collected along with the solvent and then subjected to additional analysis such as spectrophotometry or fluorometry.

gas chromatography

In gas chromatography (Fig. 12-8) separation depends on the distribution of the components between a mobile phase (an inert carrier gas) and the stationary phase. The stationary phase in this case is a nonvolatile liquid adsorbed on an inert support such as finely ground firebrick packed in a column. The sample is generally injected into the instrument as a liquid, and there it enters a heated chamber, which vaporizes it. Thus the sample actually enters the column as a vapor.

As the carrier flows through the column packed with the stationary phase, the sample (vaporized at this point) is carried along. Sepa-

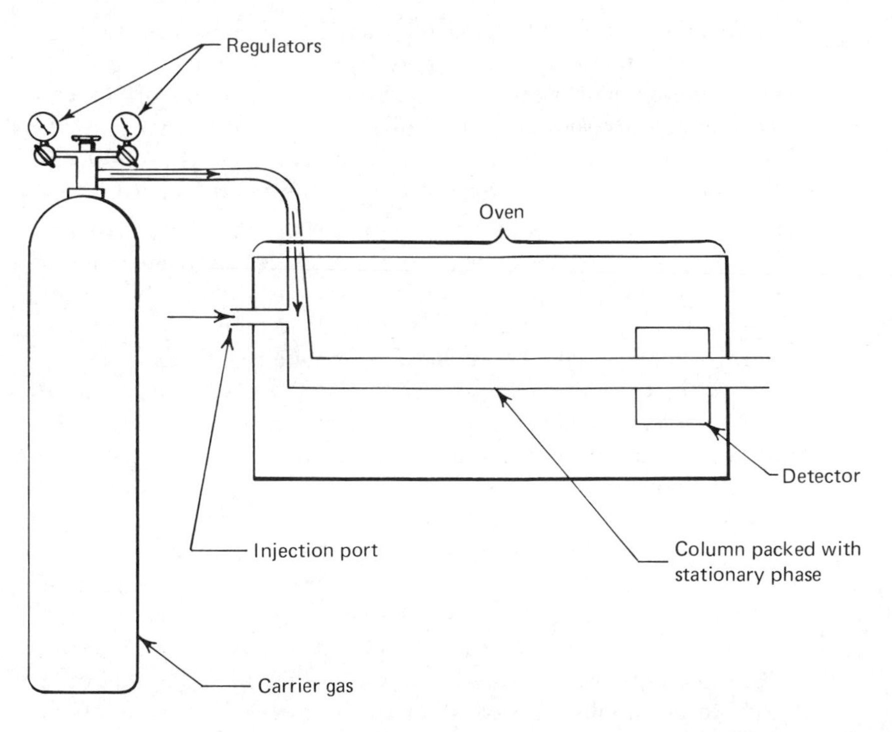

FIGURE 12–8

Schematic diagram of a gas chromatograph.

ration occurs as the vaporized constituents in the sample are distributed (partitioned) between the moving carrier gas (mobile phase) and the liquid coating on the solid support, in which the components dissolve to a greater or lesser extent dependent on their differential solubilities. Those components in the sample that have a greater affinity for the immobile liquid move through (and exit from) the column more slowly, thus effecting a separation. Chemically similar compounds travel through the column in the order of their increasing boiling points—in other words, those materials having the lowest boiling points leave the column first. In general, separation of the components is due to four basic factors: the differential solubility in the liquid coating of the solid support, the particle size of the packing material, the length of the column, and the boiling-point differences.

As the materials leave the column, they pass through a detector. When the material passes through the detector it unbalances an electrical circuit, which is then presented as a pen tracing on a chart recorder. The response of the detector varies with both the quality and quantity of material passing through, so a quantitative as well as qualitative analysis can be obtained. Three types of detectors may be employed depending on the specific analysis desired: thermal-conductivity detectors, flame-ionization detectors, and electron-capture detectors. Thermal-conductivity detectors are used in hydrocarbon analysis and electron-capture detectors are used in pesticide analysis.

Qualitative analysis with gas chromatography is based upon the amount of time that a given component is retained in the column (retention times). *Retention time* is defined as the time elapsed between the injection of a sample and the time that its exit is signaled by the detector. Identification is based on a comparison of the retention times of known samples and an unknown mixture. For example, if a sample is suspected to contain methyl, ethyl, and butyl alcohol, known pure samples of each of these alcohols would first be run through the instrument and their retention times determined. After the retention times of these samples have been determined, the unknown sample is run. These materials will separate as discussed above and pass through the detector at different intervals; thus their retention times can be determined and compared with the retention times of the knowns. By comparing retention times, the components in the sample can be qualitatively identified. In the case of questionable, very closely related materials, the sample may be collected as it passes through the detector and analyzed by additional means to obtain a positive identification.

Gas chromatography is widely used in water-management studies to determine pesticide levels, to determine the presence of various organic chemicals, and to analyze water for the presence of hydrocarbon.

pH meters and related instruments

pH and the pH scale

The acidity of a particular solution is dependent on the hydrogen-ion (H^+) concentration of that solution. The hydrogen-ion concentration of any system is based on a comparison between the hydrogen-ion concentration of the system and the hydrogen concentration of distilled water. In distilled water the water molecules are known to dissociate only very slightly, according to the equation which follows: $H-O-H \rightleftharpoons H^+ + O-H^-$. It is known that only 1×10^{-7} water molecule is dissociated (in its ionic form) at any given time. Therefore, the equation may be rewritten with the appropriate concentrations added:

$$H{-}O{-}H \rightarrow H^+ + O{-}H^-$$

$$1 \times 10^{-7} \rightarrow 1 \times 10^{-7} + 1 \times 10^{-7}$$

The common measurement of hydrogen-ion concentration is termed *pH,* which is defined as the negative logarithm of the hydrogen-ion concentration, or, mathematically, as $pH = -\log [H^+]$. Thus, in distilled water the pH would be 7, since $pH = -\log [1 \times 10^{-7}] = pH\ 7$.

Similarly, the alkalinity is dependent on the hydroxyl-ion (OH^-) concentration of a solution as compared to distilled water and is commonly termed *pOH*. pOH is defined as the negative logarithm of the hydroxyl-ion concentration or, mathematically, as $pOH = -\log [OH^-]$. Since in distilled water the dissociation of 1×10^{-7} water molecule yields 1×10^{-7} hydroxyl ion, the pOH or hydroxyl-ion concentration would be 7 because $pOH = -\log OH^-$ or $pOH = -\log 1 \times 10^{-7} = pOH\ 7$. Thus in distilled water the hydrogen-ion concentration would be 7, and the hydroxyl-ion concentration would also be 7. In other words, there would be a complete neutrality, since there would be an equal concentration of hydrogen and hydroxyl ions.

It is important to note that as the hydrogen-ion concentration increases, the hydroxyl-ion concentration must decrease. For example, if sufficient hydrogen ions are added to water to change the hydrogen-ion concentration from pH 7 ($1 \times 10^{-7}\ H^+$) to pH 6 ($1 \times 10^{-6}\ H^+$), the hydroxyl-ion concentration must decrease from $1 \times 10^{-7}\ OH^-$ to $1 \times 10^{-8}\ OH^-$. Thus, obviously, as the acidity increases, the alkalinity must decrease accordingly. It is also to be noted from the above that as the pH decreases, the hydrogen-ion concentration must increase. The situation is summarized in Table 12-1.

TABLE 12-1

acid–alkaline relationship

Acid				*Alkaline*			
H+ concentration	*pH*	*OH−*	*pOH*	*H+*	*pH*	*OH−*	*pOH*
10^{0} (1)	0	10^{-14}	14	10^{-14}	14	10^{0}	0
10^{-1} (0.1)	1	10^{-13}	13	10^{-13}	13	10^{-1}	1
10^{-2} (0.01)	2	10^{-12}	12	10^{-12}	12	10^{-2}	2
10^{-3} (0.001)	3	10^{-11}	11	10^{-11}	11	10^{-3}	3
10^{-4} (0.0001)	4	10^{-10}	10	10^{-10}	10	10^{-4}	4
10^{-7} (0.0000001)	7	10^{-7}	7	10^{-7}	7	10^{-7}	7

the pH meter

The *pH meter* measures the presence of hydrogen ions in a given sample. The basic components of a typical pH meter are the reference and sensing electrodes (Fig. 12-9) and a measuring instrument. As seen in Fig. 12-9, the *reference electrode* consists of a silver–silver chloride immersed in a saturated solution of HgCl, and both the electrode and the HgCl solution are encased in a thick glass cylinder. Since the solution surrounding the silver electrode is saturated, a constant electrical potential is established by the contact of the electrode with the ions, resulting from the normal dissociation of the HgCl. In addition, it is to be noted that the thick walls surrounding the entire apparatus are unable to conduct electrical impulses to the mercury column. Thus the potential developed at the reference electrode is constant and due solely to the solution immediately surrounding the mercury rod.

The *sensing electrode,* on the other hand, consists of a silver wire immersed in a dilute solution of HCl, all of which is surrounded by a thin membrane of special glass. Because of the characteristics of the special glass membrane, the sensitive electrode is able to detect changes in the electrical potential of the sample solution in which it is placed to measure the pH. Since the change in potential is dependent upon the hydrogen-ion concentration of the sample, the potential developed at the sensitive electrode is due to the hydrogen-ion concentration of the sample. Most pH meters manufactured today combine the sensing and reference electrodes into a single electrode.

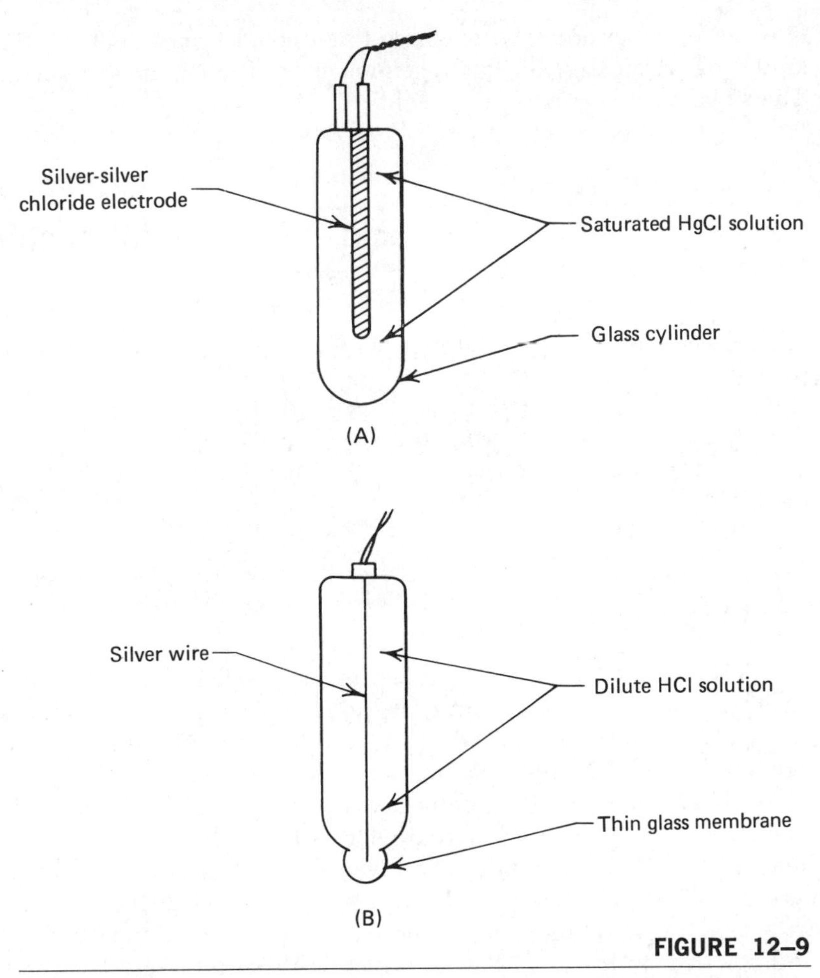

FIGURE 12–9

Reference and sensing electrodes.

The electrical potential, measured by the sensitive electrode, is compared to the stable reference potential established by the reference electrode. This current, generated by the hydrogen ions in the solution, then enters an amplifier in the measuring instrument and is used to drive a meter calibrated in pH units.

practical considerations

The pH meter must be standardized by using standard solutions, termed *buffers*, that bracket the pH of the samples to be analyzed. In other words, if the system to be measured is suspected to contain waters in the range 5 to 8, the buffers should be of pH 4 and 10. Standardization merely involves immersing the electrodes sequentially in each buffer and mechanically adjusting the pH meter to give the reading of that particular buffer. The procedure is repeated until no manual adjustment is necessary. At that time the instrument is ready for use; however, it is advisable to place the electrodes in the buffers occasionally during the course of the analysis to check the standardization.

In measuring pH it is advisable to consider the effects of temperature on pH. It is known that temperature produces a change in voltage per pH unit and also alters the dissociation of the materials in solution. Consequently, solutions will have different pH values as a result of the differences of the hydrogen-ion concentrations at various temperatures. This effect cannot be compensated for, and therefore it is not advisable to attempt to compare pH values at different temperatures. There is a temperature-compensation dial on most pH meters, and the meter can, in this manner, be adjusted to compensate for varying sample temperatures. It is, however, good practice to adjust the temperature of the buffers to within 2 degrees of the temperature of the samples when standardizing the instrument.

related instruments

Selective ion meters such as the oxygen meter operate on principles similar to the pH meter. In the oxygen meter the electrodes are combined into a single electrode. In the presence of dissolved oxygen an electrochemical reaction occurs at the twin electrode and the oxygen is converted to OH^- ions. This reaction results in the flow of current through the electrode and into the instrument, which is ultimately depicted as a meter reading. The net result of the reaction is the consumption of oxygen. The formation of OH^- ions and the production of current is, obviously, limited by the oxygen concentration in the sample and is in proportion to the concentration of oxygen present in the vicinity of the electrode. In the absence of oxygen, no reaction would occur and no current would flow. Appendix II gives the procedure for the standardization and use of a typical oxygen meter.

As noted previously, the basic principles of operation of the majority of the other specific ion meters are similar to that discussed for the pH and oxygen meter. Presently, specific ion meters and/or electrodes are available to detect ions of many elements. In addition, salinometers are available that determine in situ salinity. These instruments measure the electrical conductivity (due to the charged ions present) of marine systems. The conductivity is then automatically converted into a salinity value and given as a meter read-out. These instruments provide a rapid means whereby several components in a sample may be analyzed. They do not, however, achieve the accuracy of spectrophotometry or fluorometry.

SUGGESTED READINGS

Barnes, H. 1959. *Apparatus and Methods of Oceanography.* New York: Interscience Publishers.

Ewing, Galen W. 1969. *Instrumental Methods of Chemical Analysis.* New York: McGraw-Hill Book Company.

Skoog, Douglas A., and Donald M. West. 1969. *Fundamentals of Analytical Chemistry.* New York: Holt, Rinehart and Winston, Inc.

Strickland, J. D. H., and T. R. Parsons. 1968. *A Practical Handbook of Seawater Analysis.* Ottawa: The Queens Printer.

QUESTIONS

1 / Explain the differences between qualitative and quantitative analysis.

2 / Discuss the principles involved in spectrophotometric analysis.

3 / What are the requirements necessary for a chromogenic reagent?

4 / Explain the principles involved in visible spectrophotometry.

5 / How does atomic-absorption spectrophotometry differ from visible spectrophotometry?

6 / Explain the principles involved in fluorometry.

7 / How does spectrophotometry differ from fluorometry?

8 / Explain the operation of a gas chromatograph.

9 / Explain how retention times are used in qualitative analysis.

10 / Define pH.

APPENDIX I

general principles

part a: methods of expressing concentration

In water analysis the concentrations of the reagents used in various analysis are expressed in terms of molarity (*M*), molality (*m*), normality (*N*), and percent by volume (% vol). The concentration of the materials under investigation and which are dissolved in the water sample are customarily reported as microgram-atoms per liter (μg-at/liter), parts per billion (ppb), parts per million (ppm), parts per thousand (‰), grams per unit volume, or miligrams per unit volume. In the case of sediment or tissue analysis the amount of material contained in the actual tissues or sediments is determined and reported in parts per billion, million, or parts per thousand; at times percent by weight may also be used.

molarity

The *mole* has been defined (Chapter 1) as the gram atomic or gram molecular weight of a substance expressed in grams. In addition, it is known that 1 mole of any substance contains 6.02×10^{23} particles of that substance. Consequently, 1 mole of Na atoms would weigh 22.990 g (the atomic weight expressed in grams) and would contain 6.02×10^{23} Na atoms. Similarly, 1 mole of Cl atoms would weigh 35.453 g and contain 6.02×10^{23} Cl atoms. One mole of NaCl molecules would weigh 58.443 g (the sum of the weight of 1 mole of Na and 1 mole of Cl) and would contain 6.02×10^{23} NaCl molecules. Each NaCl molecule is composed of one Na atom and one Cl atom; thus 1 mole of NaCl would consist of 1 mole of Na and 1 mole of Cl chemically bonded to form 1 mole of NaCl.

Molarity, abbreviated *M*, is defined as moles of solute per liter of solution. In order to prepare a 1 molar (1 *M*) solution of NaCl, 1 mole or 58.443 g of NaCl would be accurately weighed out and dissolved in a minimal quantity of water. This solution would then be transferred to a 1-liter volumetric flask (Appendix I, Part B) and sufficient distilled water would be added to bring the total volume of the solution up to a 1-liter volume.

If a volume larger or smaller than a liter is required, the amount of solute and solvent are varied accordingly. For example, if only 250 ml of a 1 *M* solution of NaCl is needed, it would be uneconomical to prepare a full liter. In this case 0.25 mole or 14.6107 g of NaCl would be dissolved in 250 ml of water. This is equivalent to a concentration of 1 mole of NaCl per liter, since 0.25 mole of NaCl is dissolved in 0.25 liter of solvent. Conversely, if 2 liters of 1 *M* NaCl is required, 2 moles of NaCl (116.886 g) would be dissolved in 2 liters of water.

To prepare a solution of less than 1 *M* concentration, the concentrations of the solute would be varied accordingly. For example, if 1 liter of 0.1 *M* NaOH is required, 3.9997 g of NaOH (0.1 mole of NaOH) would be weighed out, dissolved in a minimal amount of water, transferred to a 1-liter volumetric flask, and sufficient water added to bring the volume of the solution to 1 liter. If, on the other hand, only 0.25 liter of 0.1 *M* NaOH is needed, the concentration of the NaOH must be decreased accordingly. In this case, since only 0.25 liter of solution is required, only one fourth of the amount of NaOH must be used. Therefore, 0.4444 g of NaOH would be weighed out, dissolved, added to a 0.25-liter volumetric flask, and diluted to the required volume.

In the preparation of solutions involving liquid solutes, the initial concentration of the stock solution must be known prior to the preparation of each solution. It cannot be assumed that acids such as HCl and H_2SO_4 are supplied by the manufacturer in 1 *M* concentrations. In reality, concentrated HCl generally has a molarity of 11.6 *M*, and concentrated H_2SO_4 has a concentration of 18 *M*. In preparing molar solutions of liquids, it is, therefore, necessary to know the concentration of the original material. Once this is known (generally from manufacturers' specifications), the formula $ml \times M = ml \times M$ is used. For example, if a quantity of 11.6 *M* HCl is available and it is necessary to prepare 1 liter of 0.1 *M* HCl, the required amount of 11.6 *M* HCl can be calculated according to the formula $ml \times M = ml \times M$. In this equation the left side deals with the components that must be prepared (1000 ml of 0.1 *M* HCl) and the right side considers the number of milliliters of 11.6 *M* HCl that must be used. The formula is used as follows:

Required		*To Determine*
$\text{ml} \times M$	$=$	$\text{ml} \times M$
$1000 \text{ ml} \times 0.1\ M \text{ HCl}$	$=$	$X \text{ ml} \times 11.6\ M \text{ HCl}$

In other words, the equation states the question: How many milliliters of 11.6 M HCl must be used to prepare 1 liter (1000 ml) of 0.1 M HCl? After performing the necessary calculations it would be found that 8.6 ml of 11.6 M HCl is required. In this case, therefore, 8.6 ml of 11.6 HCl would be added to a volumetric flask, and 991.4 ml of distilled water would be added to give a total volume of 1000 ml.

If less than 1 liter of solution is required, the volume portion on the right side of the equation is varied accordingly. For example, if 500 ml of 0.2 M HNO_3 (concentrated $HNO_3 = 14.8\ M$) is needed, the equation $\text{ml} \times M = \text{ml} \times M$ would be used as follows: $500 \text{ ml} \times 0.2\ M = X \text{ ml} \times 14.8\ M$. In this case it would be found that 6.7 ml of concentrated (14.8 M) HNO_3 is required.

molality

Molality, abbreviated m, is defined as moles of solute per kilogram (1000 g) of solvent. Since the vast majority of reagents in use are made up in a water solvent and since 1 ml of water weighs 1 g, there is, in reality, little difference between molality and molarity. To illustrate, if 1 kg of 1 m NaCl is to be prepared, 58.443 g of NaCl would be dissolved in 1000 g (1 kg) of water. In working with fractions of a mole or fractions of a kilogram, the amounts of solute and solvent must be varied accordingly. For example, if 0.5 kg of 1 m NaCl is required, 29.221 g of NaCl would be dissolved in 500 g of H_2O. Similarly, if 0.25 kg of 0.5 m NaCl is needed, 1.4610 g of NaCl would be dissolved in 250 g of H_2O.

normality

Normality, abbreviated N, is defined as the number of gram equivalent weights of solute per liter of solution. Normality is useful when describing the concentrations of acids and/or bases, since the strength of an acid or base is dependent upon the number of hydrogen ions (H^+) or hydroxl ions (OH^-) dissociated rather than based upon the actual amount of acid or base used.

For example, 1 liter of 1 M HCl will contain 1 mole of H^+ and

1 mole of Cl^- ions, while 1 liter of 1 *M* H_2SO_4 will contain 2 moles of H^+ and 1 mole of SO_4^{2-}. These reactions are summarized:

$$HCl \rightarrow H^+ + Cl^-$$

$$1 \text{ mole} \rightarrow 1 \text{ mole} + 1 \text{ mole}$$

$$H_2SO_4 \rightarrow 2H^+ + SO_4^{-2}$$

$$1 \text{ mole} \rightarrow 2 \text{ moles} + 1 \text{ mole}$$

Thus, although the concentrations of the acids are the same (1 *M*), the number of hydrogen ions in solution is different.

In the case of bases, the situation is similar. One liter of 1 *M* NaOH will dissociate to form 1 mole of Na^+ ions and 1 mole of OH^- ions, while 1 liter of 1 *M* $Ca(OH)_2$ will yield 2 moles of OH^- ions. These reactions are summarized by the following equations:

$$NaOH \rightarrow Na^+ + OH^-$$

$$1 \text{ mole} \rightarrow 1 \text{ mole} + 1 \text{ mole}$$

$$Ca(OH)_2 \rightarrow Ca^{+2} + 2OH^-$$

$$1 \text{ mole} \rightarrow 1 \text{ mole} + 2 \text{ moles}$$

Normality is an attempt to "compensate" for this type of dissociation and is generally considered in terms of the H^+ or OH^- ions that will be formed upon dissociation of the acids or bases under consideration. For the purposes of this text a 1 *N* solution will be considered to be that quantity of a substance that yields 1 mole of desired product upon dissociation. In the examples above involving the acids HCl and H_2SO_4 and the bases NaOH and $Ca(OH)_2$, only the hydrogen and hydroxl ions are of importance, and the other ions are customarily ignored.

Thus a 1 *N* solution of any acid will yield 1 mole of H^+ ions, and a 1 *N* solution of any base will yield 1 mole of OH^- ions when it dissociates. In the case of the monoprotic acids (one H per molecule), such as HCl, a 1 *M* solution would be identical to a 1 *N* solution; similarly, in the case of a base such as NaOH, a 1 *M* solution would be identical to a 1 *N* solution. In the cases of H_2SO_4 and $Ca(OH)_2$, however, a 1 *M* solution is equivalent to a 2*N* solution since each mole of H_2SO_4 yields 2 moles of H^+ and each mole of $Ca(OH)_2$ yields 2 moles of OH^-. These dissociations are summarized by the following equations:

$$H_2SO_4 \rightarrow 2H^+ + SO_4^{-2}$$

$$1 \text{ mole} \rightarrow 2 \text{ moles} + 1 \text{ mole}$$

$$Ca(OH)_2 \rightarrow Ca^{+2} + 2OH^-$$

$$1 \text{ mole} \rightarrow 1 \text{ mole} + 2 \text{ moles}$$

To prepare 1 liter of 1 N $Ca(OH)_2$, 0.5 mole (37.03 g) of $Ca(OH)_2$ would be dissolved in 1 liter of water since 0.5 mole of $Ca(OH)_2$ will, when it dissociates, yield 1 mole of OH^-. In the case of liquids such as H_2SO_4, the formula ml $\times N =$ ml $\times N$ is generally used to determine the amount of concentrated acid that must be used. It is to be noted, however, that the molarity of the concentrated acids must be converted to normality prior to performing the calculations. For example, assume that it is necessary to prepare 1 liter of 0.5 N H_2SO_4 and that the concentrated H_2SO_4 available is 18 M. Since each H_2SO_4 molecule dissociates into $2H^+$ ions, an 18 M solution is equivalent to a 36 N solution. The equation ml $\times N =$ ml $\times N$ is used exactly as it was in molarity calculations; in this case, however, the units are in normality. Thus

$$1000 \text{ ml} \times 0.5\ N = X \text{ ml} \times 36\ N$$

it is found that 13.8 ml of 36 N H_2SO_4 must be used. In this example, 13.8 ml of the concentrated H_2SO_4 would be added to a 1-liter volumetric flask, and sufficient distilled water would be added to give a total volume of 1 liter.

percentage concentration

Percent by weight, percent by volume, and weight-volume are commonly used methods of expressing concentrations. Frequently, the concentrations of aqueous reagents are expressed in terms of percent by weight. For example, concentrated nitric acid is said to be 70% HNO_3, which indicates that the reagent consists of 70 g of HNO_3 per 100 g of solution. Percent volume refers to milliliters of solute per 100 ml of solution, while percent weight-volume refers to grams of solute per 100 ml of solution. These methods are defined as follows:

$$\%\ \text{weight} = \frac{\text{wt of solute}}{\text{wt of solution}} \times 100$$

$$\%\ \text{volume} = \frac{\text{vol of solute}}{\text{vol of solution}} \times 100$$

$$\%\ \text{wt-vol} = \frac{\text{wt of solute (in g)}}{\text{vol of solution (in ml)}} \times 100$$

Percent weight is often used to give the concentrations of foreign materials contained in sediment or tissue samples. For example, if a 10-g sediment sample is analyzed and found to contain 0.25 g of copper, the percent weight of the copper in the sample could be calculated by using the formula

$$\frac{\text{grams (Cu)}}{\text{total sample weight}} \times 100 \quad \text{or} \quad \frac{0.25 \text{ g Cu}}{10 \text{ g}} \times 100$$

In this case, the sample would contain 2.5% by weight of Cu.

parts per million, billion

In working with very dilute solutions, percent concentration becomes awkward, owing to the number of zeros needed in front of the decimal point. In these cases the concentration is frequently expressed in terms of parts per million or parts per billion. A concentration of one part per million (1 ppm) corresponds to one part of material per million parts of the gas, liquid, or solid in which the material occurs. Parts per million and parts per billion are defined by the equations

$$\text{ppm} = \frac{\text{wt of material}}{\text{total wt of gas, liquid, or solid}} \times 1{,}000{,}000$$

$$\text{ppb} = \frac{\text{wt of material}}{\text{total wt}} \times 1{,}000{,}000{,}000$$

In salinity determinations the term parts per thousand (‰) is commonly used to express concentration. As the term implies, it refers to parts of dissolved materials per kilogram of seawater.

micrograms

In nutrient analysis involving oceanic waters, phosphate, nitrite, and nitrate are present in extremely low quantities. In these cases it has become customary to report these concentrations in terms of micrograms (μg) or microgram-atoms/liter of solution (μg-at/liter). A microgram is 10^{-6} g ($1\ \mu g = 10^{-6}$ g) or 0.000006 g.

part b: analytical glassware

The common glassware (with the exception of the familiar flasks, beakers, and test tubes) used in water analysis are volumetric flasks, pipets, and burets. To properly and accurately use any of this volumetric glassware, it is necessary to "read the meniscus" properly. When any liquid enters a tube its surface assumes a concave shape; this is termed the *meniscus*. To properly determine the exact volume of a liquid the bottom of the meniscus is located, and this indicates the true volume of the liquid (see Fig. I-1).

volumetric flasks

Volumetric flasks are available in sizes ranging from 1 ml to 5 liters. These flasks are made of this glass and have a flat bottom and narrow necks in comparison to the body. The necks are calibrated by means of an etched line, which indicates that a particular flask is able to contain a definite, stated volume of liquid. Occasionally, a second calibration mark is provided, indicating that the flask will deliver a specific volume of liquid, which is termed the "to deliver," or TD, mark. This is an important distinction, since glassware will contain a greater volume of liquid than it will deliver. This is due to the fact that a thin film of liquid will adhere to the inner walls and not leave the flask when emptied

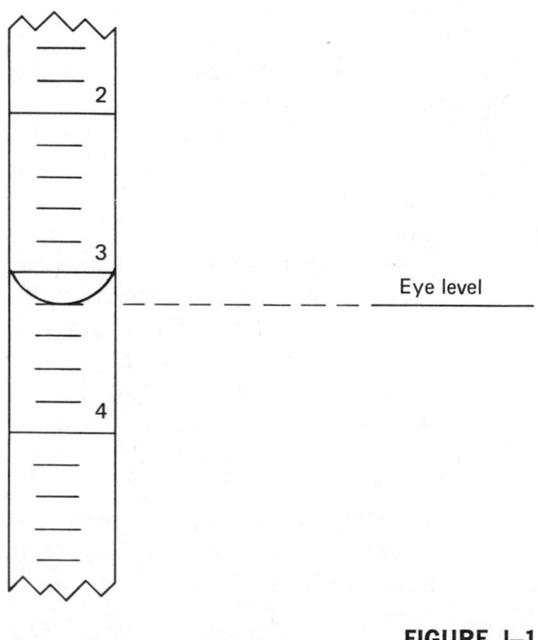

FIGURE I–1

The meniscus is read from the bottom. In this case it reads 3.2 ml.

As noted previously, the necks of these flasks are narrow in comparison to the body and contain the etched mark that the level of the liquid is brought to (the bottom of the meniscus is placed on this line). These necks are constructed in this shape so that a small error in adjusting the level of the meniscus to the calibrated line on the neck does not cause an appreciable error in the total volume of the solution.

Volumetric flasks are used in the preparation of standard solutions. In the preparation of a standard from a solid reagent, the reagent is accurately weighed and dissolved in a minimum amount of the desired solvent in a beaker. This solution is then transferred to the volumetric flask, and the beaker is washed several times with the solvent; this "wash" liquid is also transferred to the volumetric flask, and in this way it is possible to quantitatively transfer all the reagent to the flask. After all the solute has been transferred to the flask, sufficient solvent is added to the flask to bring the meniscus to the proper position on the etched calibration mark on the neck of the flask. It is to be noted that solid reagent is never added to the flask. Volumetric flasks are never used for the storage of reagents; therefore, after the solution has been prepared and adequately mixed, it is transferred to a storage bottle.

pipets

The two common types of pipets are the measuring pipet and the volumetric pipet. *Measuring pipets* have a straight, uniform bore and are generally calibrated in tenths of a milliliter. These pipets are used for the delivery of measured, variable amounts of liquid in cases where extreme accuracy is not required, since they generally have an error of approximately 1%.

Volumetric pipets, also termed *transfer pipets,* consist of a long delivery tube that is capable of reaching the bottom of a large volumetric flask, an enlarged bulb that contains the majority of the liquid, and a narrow neck with a single calibration (TD) mark. The neck is constructed similarly to the neck of a volumetric flask and for the same reason. Volumetric pipets are capable of delivering only the exact, required volume of a liquid (the volume of these pipets is indicated at the calibration mark). These pipets are available in sizes ranging from 1 ml to 100 ml and are more accurate than the measuring pipets. The deviation in delivery volume is generally less than 0.01 ml in the smaller sizes to 0.02 ml in the larger sizes.

burets

Burets are actually large, finely calibrated measuring pipets with a stopcock to control the delivery of the liquid. An efficient seal around the stopcock is achieved by greasing it with a silicon grease. Burets vary in size from 10 to 50 ml and are calibrated in tenths of a milliliter. Consequently, any volume from 50 ml or less may be delivered. Automatic burets, such as those used in salinity analysis, contain a reservoir that allows several analyses to be carried out without having to manually fill the buret.

cleaning volumetric glassware

The following cleaning solutions may be used: (1) dilute (2% or less) detergent solution or (2) chromate cleaning solution. This is prepared by adding 10 g of $Na_2Cr_2O_7$ to 200 ml of warm (100°C) concentrated H_2SO_4. After the $Ca_2Cr_2O_7$ is dissolved, the solution is cooled and placed in a glass storage bottle.

Any volumetric flask, pipet, or buret that no longer drains uniformly must be cleaned. The following procedure is recommended:

1 / Rinse thoroughly with tap water.

2 / Flush with either of the cleaning solutions (when cleaning burets, the solution should not come in contact with the stopcock, as the silicon grease will be distributed throughout the glassware by the cleaning solution). The glassware may be scrubbed with a brush if detergent solution is used.

3 / Rinse thoroughly with distilled water to remove all cleaning solution.

4 / To check the cleanliness, fill the glassware with distilled water and allow to drain. If the water drains uniformly, it is clean. If the water forms streams or droplets, the glassware must be recleaned.

part c: titrametric methods

Titration is a rapid, accurate means of determining the amount of a specific substance (solute) in solution. A titration involves the incremental addition of an exactly known standard solution to the substance under analysis. The standard solution is termed a *titrant* and is added by means of a buret.

A typical titrametric method used in seawater analysis involves the determination of chlorinity (Cl‰). The elements in seawater exhibit a constancy of composition (discussed in Chapter 6), and chlorine (in the form of chloride ions) is the most abundant element found in seawater. The chlorinity of seawater is customarily determined by titration and the salinity is then calculated from these data.

In chlorinity determinations a standard solution of silver nitrate ($AgNO_3$) is chemically reacted with the chloride ions in seawater samples. When the silver nitrate is initially prepared, it is dissolved in distilled water and, in its dissolved form, it dissociates into positively charged silver ions (Ag^+) and negatively charged nitrate ions (NO_3^-). The nitrate does not participate in any pertinent reaction during this analysis and will be ignored. The silver ions, however, react readily with the chloride ions present in the seawater sample to form an insoluble, white precipitate. This reaction is summarized by the equation $Ag^+ + NO_3^- \rightarrow AgNO_3$. It is to be noted that silver reacts with the chloride ions in a 1:1 ratio. Ideally, therefore, all that is necessary is to add incremental amounts of the standard silver solution until all the chloride is consumed (reacted). This point, again ideally, should be marked by the failure of additional silver to form any further precipitate when added to the seawater. This would indicate the absence of further chloride ions and thus mark the end of the reaction.

Since the silver nitrate solution is prepared in a manner so that each milliliter of standard solution contains one part per thousand of silver ions, and this silver reacts in a definite, known ratio with the chloride ions (the Ag^+ and Cl^- ions are said to react stoichoimetrically) in the seawater, all that is necessary, after the reaction is complete, is to determine the amount of silver solution used. This will give the parts per thousand of silver used to completely react all the chloride in the seawater. The standard silver nitrate solution is prepared in such a manner that each milliliter of standard contains 1 part per thousand of silver ions; if 15 ml of standard was used, then 15 parts per thousand of silver was used. Since silver reacts with chloride in a 1:1 ratio according to the equation $Ag^+ + Cl^- \rightarrow AgCl$, if 15 parts per thousand of silver was consumed, it must have reacted with 15 parts per thousand of chloride ions. Consequently, in this case the chlorinity would be 15‰. Because the chloride ion is in a constant ratio of Cl^- to all the other elements dissolved in seawater, all that is necessary is a simple multiplication to calculate salinity from chlorinity. The actual method involved in salinity titrations is discussed in Appendix III.

Titrations are, therefore, based on chemical reactions that may be represented by the generalized equation $aA + bB \rightarrow AB$, in which A is the titrant (silver ions in the example above), B is the substance analyzed (Cl^-), and a and b are the concentrations of each. In any titration the following requirements must be met:

1 / There should be a definite mathematical relationship between the two reactants; in other words, the reaction should be stoichiometric.

2 / The rate of reaction should be rapid.

3 / The reaction should be quantitative. For analytical accuracy the reaction must be at least 99.9% complete when a stoichiometric amount of titrant has been added.

4 / There must be a method available to determine when the reaction is complete; in other words, there must be a means of determining when to halt the addition of further standard solution.

Using a typical salinity titration as an example, it was noted previously that all that is ideally necessary is the addition of the standard silver nitrate solution until the precipitate no longer forms. Unfortunately, in this reaction once the first drop of standard is added to the reaction flask, the resultant solution becomes so cloudy that it is impossible to determine at what point the precipitate ceases its formation. In the case of salinity analysis, this problem is solved by the addition of a chemical indicator, termed the *chromate ion* ($CrO_4^=$), to the seawater sample prior to analysis. Consequently, the standard silver nitrate solution is added to a seawater sample which also contains a chromate indicator. Since the chromate is negatively charged, it can be expected that the positively charged silver ions will react with the negatively charged chloride ions as well as with the chromate ions. In reality there is a "preferential affinity" between the silver and the chloride. In other words, the silver will react preferentially with the chloride ions until all the chloride is utilized (converted into AgCl). Only after the chloride has been utilized will the silver ions permanently react with the chromate ions to form silver chromate ($AgCrO_4$). The formation of silver chromate is marked by the solution taking on a definite pinkish tinge. Since the function of the chromate is to indicate when the reaction is complete, it is termed a *chemical indicator,* and the point at which further addition of standard should be terminated is termed the *end point.* The total reaction may be summarized by the following equations:

	(Step 1)	$+ Cl^-$ (in solution)	$\rightarrow$	$AgCl\downarrow$ (white ppt)
Ag^+				
	(Step 2)	$+ CrO_4^=$ (indicator)	$\rightarrow$	$AgCrO_4$ pink (end point)

Standard solutions used in both titrations and in various methods of spectrophotometric analysis are generally prepared by accurately weighing out the desired amount of a highly purified form of the appropriate chemical (termed a *primary standard*). This material is then dissolved in distilled water, transferred to a volumetric flask, and quantitatively diluted to the proper volume. A reagent must fulfill several requirements if it is to be used as a primary standard: (1) the material must be of exactly known composition and highly pure; (2) it should react rapidly and stoichiometrically with the material being analyzed; and (3) the standard should remain stable at room temperature, withstand drying in an oven, and not decompose when exposed to light. In addition, it should not absorb water and other materials from the atmosphere.

Two major types of titration are commonly used in water analysis: precipitate-forming titrations (discussed above), and titrations involving the formation of chemical complexes, generally used in metal analysis. Most positively charged metal ions can be determined by titration with standard solutions of an organic complexing agent such as ethylenediaminetetraacetic acid (EDTA). EDTA reacts with most metals to form a stable water-soluble complex. The reactant ratio of EDTA to metal ion is generally always 1:1. An indicator is added prior to the titration, which forms a highly colored complex with the metal ion. Since the preferential reaction in this case is between the metal and the EDTA, the color imparted by the indicator–metal complex will be visible only until sufficient EDTA has been added to consume all the metal. As the EDTA reacts with the metal, the metal is released from the indicator and the color fades. The EDTA determination of water hardness (Chapter 8) is given in Appendix III. Generally, trace amounts of metals are determined by atomic-absorption spectrophotometry.

part d: grades of laboratory reagents

This list is in order of decreasing purity.

1 / Primary Standards: reagents of high-precision assay. These reagents are suitable for use in the preparation of standard solutions.

2 / Electronic Grade Solvents: dry solvents containing less than 0.1 ppm metallic impurities.

3 / Spectranalyzed Reagents: spectrophotometrically pure and supplied with individual absorption curves.
4 / Certified Reagents: meet the American Chemical Society standards for high purity.
5 / USP: made to the specifications of the *U.S. Pharmacopeia.*
6 / NF: made to the specifications of the *National Formulary.*
7 / Highest Purity: organic chemicals whose purity is determined by melting or boiling point or by refractive index.
8 / Purified: free of extraneous foreign matter and suitable for use in synthesis.
9 / Technical Grade: these chemicals are physically clean and of reasonable chemical purity.

part e: standard solutions

The majority of the analytical methods utilized in water analysis are volumetric in nature and involve the addition of a solution of known concentration, termed the *standard solution,* to a solution of unknown concentration, termed the *sample.* The addition continues until chemically equivalent quantities of the two solutions have been brought together. At this point the volume of the standard used is employed to determine the quantity of material present in the same. These methods, termed *titrametric methods,* are commonly used in determinations of salinity, dissolved oxygen, and biological and chemical oxygen demand.

The analyst relies on a physical change in the solutions to indicate when the reaction is complete. Usually, the physical change is in the form of a color change of a material, termed an *indicator,* which is added to the sample for this specific purpose. The following solutions are of importance in chemical analysis.

A *standard solution* is one of precisely known concentration. As the name implies, the standard solution is one against which comparisons of the strengths of the components of samples are made. There are two methods of preparing standard solutions, and both depend upon the use of a primary standard.

A *primary standard* is a substance of the highest purity. It can be dried, weighed accurately on an analytical balance, and is known to react in a definite, predictable ratio with the desired material. Primary standards are prepared as follows. Reagents of the proper purity are selected, dried, and the desired quantity accurately weighed (to the fourth decimal place). The material is then dissolved in the minimum quantity of distilled water, quantitatively transferred to a volumetric flask of the desired volume, and distilled water added to the mark. The concentration of this solution is accurately known, since an exactly known weight of reagent is added to an exactly known volume of solvent. Primary standards are available for practically every purpose. All primary standards must be of the highest purity and be capable of (1) being dried at temperatures between 100 and 110°C, (2) being weighed with no more than ordinary care and precautions, and (3) reacting in a known and predictable ratio with the desired material (react stoichiometrically).

This method of standard preparation makes use of a large quantity of rather expensive primary standard. To avoid excessive use of the primary standard, a *secondary standard* is generally prepared, standardized against the primary standard, and then the secondary standard is generally used in the analysis. Since the secondary standard is generally prepared from less costly materials, this method effectively keeps costs at a reasonable level. To prepare a secondary standard, the suitable reagent is weighed, dissolved in distilled water, quantitatively transferred to a suitable volumetric flask, and distilled water added to the mark. The concentration of this solution is then determined by titration with the proper primary standard.

APPENDIX II

use of analytical instruments

part a: the analytical balance

The determination of the weights of both standard reagents and samples is the most basic and critical laboratory operation encountered. Without proper weight determinations the analysis will be inaccurate, regardless of the care that the remainder of the analysis receives. Several types of *analytical balances* are available. However, most commercial laboratories now use *single-pan, direct-reading electric balances,* and only this type will be discussed here. These balances are capable of rapidly weighing materials to the third or fourth decimal place with a minimum amount of skill and maintenance (Fig. II-1).

In electric balances of this type a beam is mounted from front to back in the instrument case. A counterweight, which functions as the familiar second pan, is mounted on the rear portion of the beam. This counterweight is balanced by a series of movable weights (mounted in the case) along the front of the beam. The movable weights are removed from the beam by levers controlled by knobs mounted on the front of the case. When the balance is unloaded (no weight on the pan), the counterweight is exactly balanced by the movable weights and the balance is said to be *zeroed.* The zero may be manually adjusted, if necessary, by a knob located on the side of the instrument case.

When the material to be weighed is placed on the pan, the instrument becomes "unbalanced." In other words, the sum of the movable weights and the substance to be weighed are greater than the weight of the counterweight. The weights are equalized in the weighing process by removing the movable weights from the beam. Thus these single-pan balances weigh by substitution.

The movable weights are removed from or added to the beam by means of manipulating a series of knobs located on the face of the instrument case. These knobs are connected to a series of levers which

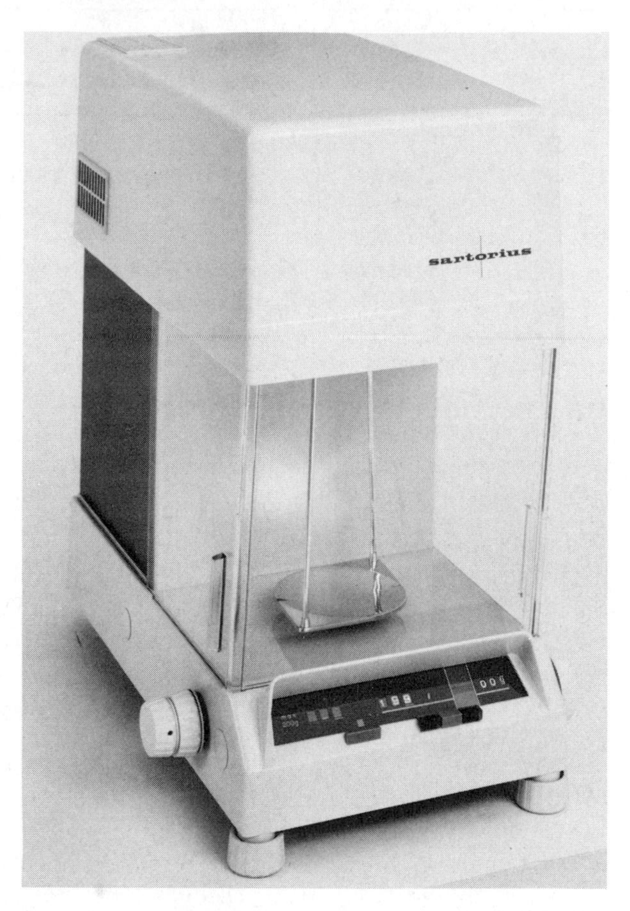

FIGURE II–1

The single-pan analytical balance. (*Courtesy The Sartorius Balance Division of Brinkman Instruments, Inc.*)

pick up the movable weights and remove them from the beam. After the proper number of these weights have been removed, the counterweight is exactly balanced by the weight in the sample on the single pan. There are generally three knobs on the face of the instrument that serve to remove the movable weights. One will remove from 10 to 150 g (removes tens of grams) of weight, one removes from 1 to 9 g (removes grams), and one removes tenths of a gram from 0.1 to 0.9 g. An illuminated scale on the front panel gives the last two or three digits. As noted previously, the balance is zeroed by means of a knob on the side of the case, and a lever on the other side can assume three positions. One position locks the beam and prevents any movement; one position is a partial unlock, which allows restricted beam movement; and the third position is a complete unlock, which allows full and unrestricted movement of the beam and the pan. Weights are removed from the beam in the partial-unlock position only.

The single-pan balance is operated in the following manner:

1 / With no weight on the single pan, the balance is placed in the complete unlock position and the zero is checked (on the illuminated scale). If necessary, the balance is zeroed by the zero knob.

2 / The balance is locked and the material to be weighed is placed on the pan. The material is *never* placed directly on the pan but is weighed in a preweighed weighing bottle.

3 / The balance is placed in the partial-unlock position and the appropriate weights are removed from the beam. As the knobs remove the weights, the digits appear on the face of the case.

4 / After sufficient weight has been removed, the balance is placed in the complete-unlock position. At this point the last two or three digits are given on the illuminated scale. The weights indicated are recorded.

5 / The balance is locked, the movable weights are placed on the beam, and the material is removed from the pan.

part b: the pH meter

The theory of the operation of the pH meter has been discussed in Chapter 11. These instruments may be either line or battery-operated, consist of a reference and a sensing electrode or a single combination electrode, and some models may be designed for use with a chart recorder. Regardless of the power source and instrument design, all pH meters encompass the following operational components: a temperature compensator (a dial to manually adjust the instrument to read at the temperature of the samples being analyzed), a standardized dial which manually adjusts the meter dial, and an on–off–standby switch. The "standby position" is termed the "reference position" on some models. The pH meter is operated as follows:

1 / The instrument is turned to the standby position and allowed to warm up. Suggested times are given in the manufacturer's specifications.

2 / To standardize the instrument, take the temperature of the appropriate buffers and set the temperature compensator dial accordingly.

3 / Place the electrodes into the buffer of the lower pH; turn the meter on and manually adjust the meter (using the standardize control) to the pH of the buffer.

4 / Turn the switch to standby; remove the electrodes from the buffer and, using a wash bottle, rinse the electrodes thoroughly with distilled water.

5 / Place the electrodes in the second buffer; turn the instrument on and manually adjust the meter to the pH of this buffer.

6 / Turn the meter to standby; remove the electrodes from the buffer and rinse.

7 / Repeat steps 2, 3, and 4 until no manual adjustment is necessary to obtain the pH of each buffer.

8 / Turn the meter to the standby position, remove the electrodes from the buffer, and rinse.

9 / Place the electrodes into the sample; switch the instrument on and record the pH of the sample.

10 / After each pH determination, the meter is placed on standby prior to removing the electrodes from the solution and rinsing.

11 / After all the samples have been analyzed, the instrument is turned off, the electrodes removed from the solution, and rinsed. The electrodes are stored in distilled water when not in use.

It is to be noted that the meter is always placed in the standby position before the electrodes are removed from the solution.

part c: spectrophotometry

The most common spectrophotometer in use is the Bausch and Lomb Spectronic 20®, and the directions for instrument use given here will be based upon the operation of the Spectronic 20®. In addition, directions are given for a student standardization and determination of an unknown. Appendix III will give actual spectrophotometric methods of water analysis.

Figure II-2 shows the instrument. Located in the center of the instrument case is the meter, which can give readings in either percent transmission or absorbance (see Chapter 11). The knob on the

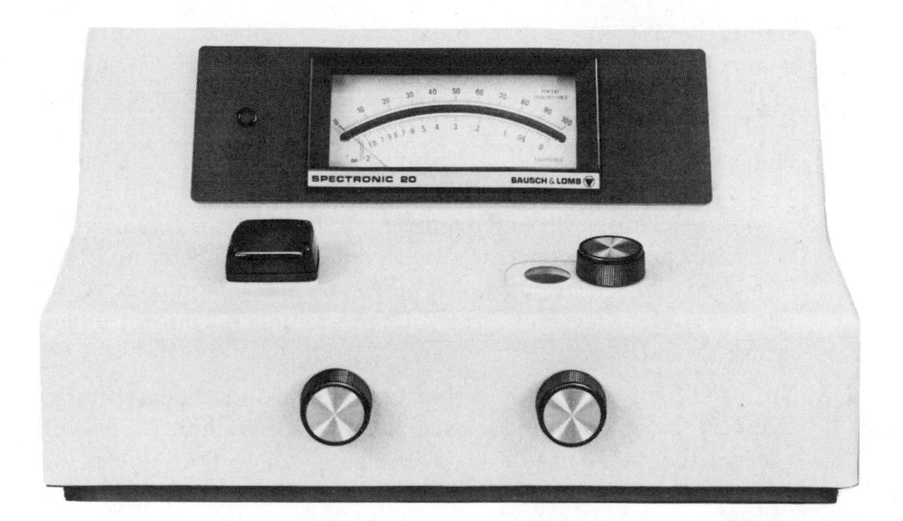

FIGURE II–2

The Spectronic® 20. (*Courtesy Bausch and Lomb, Analytical Systems Division*)

right is the wavelength selector, which is calibrated in 5-mμ divisions. This moves the monochrometer, thus passing the proper light through the sample. To the left is the sample holder and cover and the "zero" control (see Chapter 11 for a discussion of blanking and zeroing a spectrophotometer). To the right of the meter on the newer instruments is a red light that glows when the instrument is turned on. The newer instruments have a built-in voltage regulator and may be plugged directly into the line. The older models (without the red on–off light) require an external voltage regulator. In these models the Spectronic 20® is plugged into the voltage regulator and the regulator is plugged into the line. The instrument is operated as follows:

1 / The instrument is turned on by rotating the left-hand knob clockwise. The instrument should be allowed to warm up for 10 to 15 minutes.

2 / Using the wavelength selector (right-hand knob), select the proper wavelength.

3 / After the instrument has warmed up, and with no sample in the sample holder, the instrument is zeroed. This is accomplished by using the left-hand zero-adjust knob and placing the meter on 0% transmittance (infinite absorbance). If the zero reading is not stable, the instrument must be allowed further warm-up time. It is important to note that the instrument is zeroed with no cuvette in the sample holder. This is due to the fact that when a cuvette is placed in the sample holder, it automatically removes an occluder. When the occluder is in position (with no cuvette in the instrument), it blocks the light and prevents it from hitting the phototube. Thus, with no cuvette in the instrument, the occluder is in place and no light is striking the phototube. Therefore, there is a zero light transmission to the phototube and conversely maximum absorbance of light (by the occluder). The instrument is now zeroed.

4 / Turn the light-control knob (located on the right side of the instrument case) counterclockwise *almost* to its limit. This will prevent the meter needle from "pegging" when the light passes to the phototube.

5 / Insert the cuvette containing the blank (see Chapter 11) into the sample holder. As the cuvette enters the holder, it will automatically remove the occluder from the light path, allowing the light to strike the phototube. After the cuvette is in place, the sample-holder top must be closed. The light-control knob is now adjusted so that the meter reads 100% transmittance (zero absorbance). The instrument is now blanked.

6 / Recheck the zero and blank and readjust if necessary.

7 / During any standardization and analysis, the zero and blank should be checked frequently and readjusted if necessary.

8 / The instrument may now be standardized, and following this operation, unknowns may be analyzed.

It is to be noted that the cuvettes must only be handled by the upper portions. This is necessary since the light passes through the midregions, and handling this portion will place fingermarks on the walls, which will interfere with the light transmission through the cuvette. To avoid this, it is good practice to clean the cuvettes with a cleaning tissue (lens paper) prior to inserting them into the sample holder.

A typical standardization and quantitative analysis is given below. It is to be noted that the solution ($Na_2Cr_2O_7$) is highly colored, and therefore the chromogenic development of color is not necessary.

In this analysis the exact concentration of a $Na_2Cr_2O_7$ solution will be determined. Since the wavelength of maximum absorbance is not given, it will be necessary to determine the proper wavelength by preparing an absorbance spectra (see Chapter 11) prior to either the standardization of the instrument or the quantitative analysis of the unknown. The suggested analytical method follows:

1 / Since the wavelength is unknown, the instrument should be zeroed and blanked at an arbitrary wavelength. The blank in this case is distilled water, since it is the solvent in which the $Na_2Cr_2O_7$ is dissolved. The instrument is blanked and zeroed as discussed previously.

2 / Since no color development is necessary in this case, the sample may be placed in the sample holder with no prior preparation.

3 / After the sample is placed in the sample holder, the wavelength is set to 400 μm and the absorbance is recorded. The wavelength is then taken through the entire spectrum in increments of 10 μm. The absorbance is recorded at each wavelength.

4 / After all the data have been obtained, a graph of wavelengths versus absorbance is prepared.

5 / From this absorbance spectra, it will be noted that maximum absorbance occurs only in a specific region (at a specific wavelength).

6 / The wavelength of maximum absorbance is determined and the instrument set at this wavelength.

Standardization: preparation of a standard curve

1 / A stock solution of 5.0000 g/liter of $Na_2Cr_2O_7$ is prepared (this is equivalent to 5 mg/ml) in a volumetric flask and is transferred to a storage bottle.

2 / By a series of accurate dilutions (using volumetric pipets and volumetric flasks), prepare 100 ml of the following $Na_2Cr_2O_7$ solutions: 2.50 mg/ml, 1.25 mg/ml, 0.625 mg/ml, 0.312 mg/ml, 0.156 mg/ml, 0.078 mg/ml, and 0.039 mg/ml. These solutions are the standards.

3 / Blank and zero the instrument.

4 / Place portions of each of the standards into cuvettes and determine the absorbance of each.

5 / Prepare a curve of concentration versus absorbance. This is the standard $Na_2Cr_2O_7$ curve and may be used to determine the concentration of $Na_2Cr_2O_7$ unknowns.

6 / The instrument should be zeroed and blanked if necessary and the unknowns analyzed.

part d: atomic-absorption spectrophotometer

The atomic-absorption spectrophotometer is used in the analysis of elements in either water samples or in sediment or tissue. In the case of dissolved elements, the water sample may generally be analyzed directly, while in the analysis of sediments or tissue the elements must be extracted from the sample prior to analysis. Extraction methods are given in Appendix III.

The atomic-absorption spectrophotometer has the following operational controls: a wavelength selector, which adjusts the position of the monochrometer and allows it to pass only the specific light emitted by the lamp on to the detector; a meter, to indicate the amount of this light absorbed by the sample; a minimum and maximum adjustment control; and knobs to control the flow of the acetylene and air that provides the fuel for the flame. The instrument is operated as follows:

1 / The proper lamp is placed in the lamp compartment, aligned, and the instrument is turned on and allowed to warm up.

2 / After the warm-up period with the lamp on, aligned, and emitting the proper wavelength, the monochrometer is set. This is accomplished by slowly and carefully rotating the wavelength selector while watching the absorbance meter. The monochrometer (by means of the wavelength selector) is adjusted so that a minimum absorbance reading is obtained. It is not necessary to go through all the possible monochrometer settings to accomplish this, since the manufacturer's specifications will list the recommended approximate wavelengths to be used for each element. Thus the monochrometer is set approximately according to the specifications. After the approximate setting the wavelength selector is varied over a relatively narrow range, while noting the meter reading until the proper wavelength is located.

3 / The air–acetylene mixture is set, the valve opened, and the flame lit.

4 / The instrument is zeroed by aspirating the proper solvent (solvent with no sample) and adjusting the meter to give a zero-absorbance reading with the minimum concentration control.

5 / A series of standard solutions are available (or prepared in a similar manner to the standards discussed in Part C). The most highly concentrated standard is aspirated, and the meter is placed on maximum absorbance by adjusting the maximum concentration control.

6 / Distilled water should be aspirated at this time to flush any traces of the standard. The zero and maximum readings should be rechecked and the instrument readjusted if necessary.

7 / The prepared standards are aspirated sequentially, and their absorbance is recorded. It is to be noted that it is good practice to flush the aspirator with distilled water after each standard is aspirated to avoid false readings due to the presence of the previous standard.

8 / A standard curve of concentration versus absorbance is prepared.

9 / Unknown samples are aspirated, their absorbance determined, and the concentration calculated from the standard curve. As noted previously (step 7), distilled water should be aspirated after each sample to prevent contamination. In addition, the zero and maximum concentration should be checked frequently and the instrument readjusted if necessary.

part e: oxygen meters

Although there are several models of oxygen meters available, they all employ either a silver–platinum or gold–platinum electrode, and all operate according to similar principles (the development of an electrical potential at the electrode due to the oxygen concentration as discussed in Chapter 11). These instruments may be either line-operated for use in the laboratory or aboard large vessels or battery-operated for use in the field or aboard small vessels.

The mode of power supply will vary depending upon the type of meter employed; however, all meters generally have the following controls in common: a meter calibrated in oxygen (generally ppm) and in temperature (in degrees centigrade), a battery check switch (on the field models), a "zero" control, and a "read" switch.

The battery check switch, as the name implies, is used prior to beginning the analysis to ensure that the battery is charged. The zero control is used to manually set the oxygen reading on the meter to zero. The temperature switch automatically switches the function of the meter to read in terms of degrees centigrade. Generally a thermistor (temperature-sensitive wire) is attached to the meter, and in this manner temperature is determined. The read switch will electrically cause the meter to read the oxygen concentration of the solution in which the probe is emersed. Generally, this switch has two positions, which serve to change the scale reading on the meter. One scale is capable of giving readings in the range 0–10 ppm O_2, while the other scale reads the 0–20 ppm range. Thus it is possible to expand the scale (by means of the switch setting) in the lower portion (0–10) to obtain a more accurate reading.

The meter is either air-calibrated to the amount of oxygen contained in air at a given temperature or to the amount of O_2 in a saturated solution of water at a given temperature. In either case the amount of oxygen capable of being held in solution (this amount in air or water at that temperature is known and given with the operating instructions), and the meter can be set manually to this reading prior to beginning the analysis.

Oxygen meters, when properly standardized and used, give a more rapid and accurate analysis than the titrametric methods discussed in Appendix III. It is recommended that the manufacturer's instructions for the oxygen meter being utilized be consulted prior to use.

part f: salinometers

Salinometers operate in a manner similar to pH meters (Chapter 11). In the case of salinometers, however, the electrical conductivity developed by the total dissociated, dissolved materials in seawater is measured by the probe. The electrical potential is then converted to salinity internally by the instrument and given as a meter read-out in parts per thousand (‰). This conversion is valid due to the constancy of composition of seawater (Chapter 6).

Although salinometer design varies greatly, they have the same basic mode of operation. The operation of the Beckman induction salinometer is as follows. The instrument consists of a probe that is lowered into the water column to measure the conductivity, an on–off switch, and a four-place switch that in its various positions can be used to check the battery (battery check), the conductivity, the salinity, and the temperature (since there is a thermistor in the probe). There are two slots on the face of the meter, a conductivity–temperature slot and a salinity slot.

The salinometer is manually calibrated initially by immersing the probe in a seawater sample of known salinity (determined by titration; see Appendix III, Part A). Thereafter, it may be calibrated by placing a known potential (in the form of an electrical resistor) on the probe, switching to the conductivity setting, and checking the reading.

If a suitable reading is obtained, the meter may be used. However, if the conductivity does not give the recommended reading, the meter must be recalibrated in standardized seawater.

To use the instrument, the following procedure is used:

1 / Prior to setting out, switch the meter on. Place the four-place switch to battery check. If the meter reads "in the green" (on the green portion of the meter), the battery is charged.

2 / Switch the meter off.

3 / Upon reaching the station, immerse the probe and switch the meter on.

4 / Place the four-way switch to "conductivity." The meter will deflect from the zero position. Using the dial opposite the conductivity/temperature slot manually, adjust the meter to read zero. As the adjustment is carried out, the numbers within the slot will change. When the meter reads zero, note the numbers within the conductivity slot. This is the conductivity of the water.

5 / Place the switch to salinity. The meter will deflect from the zero position. Using the dial opposite the salinity slot, manually adjust the meter to read zero. The numbers within the slot will change as the adjustment is performed. After the meter reads zero, take the reading of the numbers within the slot. This is the salinity.

6 / Place the switch on temperature. The meter will again deflect from zero. Using the dial opposite the conductivity/temperature slot, adjust the meter to read zero and note the reading. This is the temperature of the water.

7 / After all the data have been taken, the meter is switched to off. The probe and face of the instrument should be washed in fresh water prior to storage.

APPENDIX III

water-analysis methods

This appendix discusses the more common analytical methods used in water analysis. After the samples have been collected, they must be preserved if the analysis is to be delayed. The EPA guidelines for the preservation of samples, as well as the recommended holding times, are given in Table III-1.

TABLE III-1

methods of sample preservation

Analysis	*Preservative*	*Holding Time*
pH	Determine on site, or hold at 4°C	6 hours
Heavy metals	Adjust to pH 2 with HNO_3	6 months
BOD	Hold at 4°C	6 hours
COD	Adjust to pH 2 with H_2SO_4	7 days
Chloride	Filter on site; hold at 4°C	7 days
Dissolved oxygen	Determine on site, or see p. 221	
Nitrate; nitrite	Filter on site; adjust to pH 2 with H_2SO_4; hold at 4°C	24 hours
Phosphate	Filter on site; hold at 4°C	24 hours
Turbidity	Hold at 4°C	7 days

part a: salinity analysis

Salinity is determined by analyzing a given water sample for the chloride concentration by titration with silver nitrate; from these data, the salinity is calculated. Because of the constancy of composition of seawater (Chapter 6), it is possible to calculate salinity from chlorinity.

It is necessary to standardize the silver nitrate since all laboratories that perform salinity determinations similarly standardize their $AgNO_3$. This allows the laboratories to perform the analysis with a standard concentration of $AgNO_3$, thus obtaining uniform results among all laboratories.

It is known that the silver nitrate reacts stoichiometrically in an exact 1:1 ratio with chloride ions. Therefore, ideally if the silver nitrate solution is prepared properly, exactly 1 ml of $AgNO_3$ should react completely with 1 ml of standard, 10 ml of $AgNO_3$ should react completely with exactly 10 ml of the standard, and so on. If, however, the $AgNO_3$ is too dilute, more than 10 ml of $AgNO_3$ will be required to react completely with exactly 10 ml of the standard. Conversely, if the $AgNO_3$ is too concentrated, the opposite will occur.

Rather than directly correct the concentration of the $AgNO_3$, it is customary to obtain a correction factor, designated alpha (α), on the basis of a titration against standard seawater. If the $AgNO_3$ is too concentrated, it will take less than 10 ml to reach the end point, the α will be assigned a negative sign, and in all calculations involving this α it must be subtracted. If the $AgNO_3$ is too dilute, it will take more than 10 ml to reach the end point, and the α will be given a positive sign. Thus in all calculations involving this α, it must be added.

After the $AgNO_3$ has been standardized, the samples may be analyzed and the chlorinity determined. The salinity may then be calculated by use of conversion tables or by means of the formula $S‰ = 1.805\ (Cl‰ \pm \alpha) + 0.03$, in which Cl‰ is obtained directly from the buret reading (see Appendix I, Part C).

The procedure is as follows. It is recommended that an automatic 10-ml pipet and an automatic zero-adjusting 25-ml buret be used. The titration is to be carried out in a 250-ml beaker or Erlenmeyer flask.

Reagents

1 / Standard seawater (Copenhagen water).

2 / Silver nitrate solution (0.28 *N*): dissolve 49.0 g of reagent-grade $AgNO_3$ in distilled water, transfer to a 1-liter volumetric flask, and dilute to the mark. If a large number of samples are to be analyzed, it is recommended that several liters of $AgNO_3$ be prepared and standardized.

3 / Chromate indicator: dissolve 3.5 g of reagent-grade potassium chromate (K_2CrO_4) in distilled water, transfer to a 1-liter volumetric flask, and dilute to the mark.

Silver nitrate standardization

1 / Since each lot of Copenhagen water varies slightly in its stated chlorinity (the chlorinity is precisely known and stated, but it varies with each lot), it is necessary to calculate the number of milliliters of $AgNO_3$ required to reach an end point with that particular lot of Copenhagen water. This is done by assuming that the $AgNO_3$ has been prepared exactly and then performing the calculation; for example if the stated Cl‰ of the standard seawater is 19‰ and the $AgNO_3$ has been prepared to be 19‰, the required number of milliliters of $AgNO_3$ needed to reach the end point is calculated by the formula ml × Cl‰ (of standard) = ml × Ag‰ (of silver nitrate). This is merely a variation of the equation used to calculate N and M given in Appendix I, Part A. This calculation will give the number of milliliters of $AgNO_3$ required to reach an end point, assuming that the $AgNO_3$ was prepared precisely. The deviation from this number will give the α factor.

2 / Place exactly 10.00 ml of the Copenhagen water in a 250-ml beaker, and add 15 ml of the potassium chromate indicator.

3 / Place the $AgNO_3$ in the buret and perform the titration. The end point is marked by the formation of a persistent faint brownish-red color.

4 / The titration should be repeated five times, the results averaged, and the α determined.

5 / The α should be noted and placed on the stock $AgNO_3$ storage bottle.

Analytical procedure

1 / Add 10.0 ml of sample to the beaker, add 15 ml of the indicator, and titrate with the standardized $AgNO_3$ to the proper end point.

2 / Note volume of the $AgNO_3$ added (this is approximately equal to the Cl‰).

Calculations

1 / The S‰ may be calculated from the Cl‰ by use of conversion tables, or

2 / By means of the formula S‰ = 1.805 (Cl‰ $\pm$ α) + 0.03.

NOTE—This method can also be used to trace the movement of groundwater (Chapter 5). Generally, however, quantitative data are not necessary, and the chloride (introduced in the recharge well as NaCl or KCl) can be qualitatively determined at the test wells by ob-

taining water samples, adding 1 or 2 ml $AgNO_3$, and checking for the formation of the AgCl precipitate. By determining the distance of the recharge from the test well and the time elapsed from the introduction of the Cl ion, the rate of flow can be determined.

part b: oxygen analysis

There are three analytical methods that may be used in the determination of oxygen levels: *dissolved oxygen analysis* (*DO*), *biological oxygen demand* (*BOD*), and *chemical oxygen demand* (*COD*). The choice of analysis is dependent upon the data required and the origin of the water sample under analysis. For example, DO is commonly performed on samples taken from natural systems. The BOD is a measure of the oxygen removed from systems over a given period of time by biological activity in the course of the decomposition of organic material. BODs are commonly performed in areas where there is a known or suspected input of organic material. The COD is a measure of the oxygen demand placed on a given system by chemical input (chemical oxidants). This method would be performed on waters receiving, or suspected of receiving, this type of input. The methods for DO, BOD, and COD analysis follow.

dissolved oxygen (standard methods): method I

In the determination of dissolved oxygen, various ions and compounds interfere with the analysis. Consequently, there have been several methods devised to correct for these interferences. The method given below, termed the *azide modification*, most effectively removes the interference due to nitrite. This is a titrametric method in which the sample, following various preparatory steps, is titrated with a standard solution of sodium thiosulfate to a pale-straw-colored end point.

Reagents

1 / Manganese sulfate solution: dissolve 480 g of $MnSO_4 \cdot 4H_2O$ in distilled water, filter, transfer to a 1-liter volumetric flask, and dilute to the mark.

2 / Alkali iodide–azide reagent: dissolve 500 g of NaOH and 135 g of NaI in distilled water, transfer to a 1-liter volumetric flask, and dilute to the mark.

3 / Dissolve 10 g of NaN_3 in 40 ml of distilled water and add to the NaOH–NaI solution. When this solution is added to the starch solution (reagent 5), diluted, and acidified, it should not give a color.

4 / Concentrated H_2SO_4.

5 / Starch solution: add 6 g of soluble starch to a small quantity of distilled water and mix (it will not dissolve at this point). Add the starch mixture to 1 liter of boiling water and allow to boil for 5 minutes and stand overnight. Use the clear supernate in the analysis. This solution may be preserved with 1.25 g of salicylic acid or 2 drops of toluene per liter.

6 / Sodium thiosulfate stock solution (0.10 *N*): dissolve 24.82 g of $Na_2S_2O_3 \cdot 5H_2O$ in boiled and cooled distilled water, transfer to a 1-liter volumetric flask, and dilute to the mark. This solution is preserved by the addition of 5 ml of chloroform.

7 / Standard thiosulfate titrant (0.025 *N*): quantitatively transfer 250.0 ml of the stock solution (reagent 6) to a 1-liter volumetric and dilute to the mark with freshly boiled and cooled distilled water. This solution, if prepared accurately, will have a normality of 0.025 and be equivalent to 0.200 mg of DO/1 ml. This solution is preserved by adding 5 ml of chloroform per liter of solution.

8 / Standard potassium dichromate solution (0.025 *N*): dissolve 1.226 g of $K_2Cr_2O_7$ in distilled water, transfer to a 1-liter volumetric flask, and dilute to the mark. The 0.025 *N* $K_2Cr_2O_7$ solution is used to standardize the 0.025 *N* sodium thiosulfate by the following procedure.

Standardization of the thiosulfate / Dissolve 2 g of KI in 150 ml of distilled water and add 10 ml of H_2SO_4, followed by exactly 20.00 ml of the $K_2Cr_2O_7$ standard. Dilute to 200 ml and titrate with the thiosulfate standard. When a pale straw color is achieved (toward the end of the titration), add 2 ml of the starch solution. The addition of the starch will be marked by the formation of a blue color (the iodine–starch reaction). Continue the addition of thiosulfate to the first disappearance of the blue color. Assuming that the thiosulfate is exactly 0.025 *N*, precisely 20.00 ml of thiosulfate should be required to reach the end point. If necessary, it is convenient to adjust the concentration of the thiosulfate (on the basis of the standardization) to 0.025 *N* to avoid additional calculations.

Sampling and analytical procedure / All water samples should be collected in BOD bottles. These bottles have a capacity of either 250 or 300 ml and are sealed with a ground-glass stopper that prevents the addition of atmospheric air. When the sample is collected, the bottle should be "overfilled" with the sample and the stopper promptly inserted. At this time some of the sample will overflow from the bottle. If this method is followed, no air will be trapped inside the bottle. It is necessary to eliminate air from the BOD bottle in this manner, since trapped air will tend to diffuse into the sample, giving erroneous results. When reagents are added to the sample (in the BOD bottle), the pipet is placed in the bottom of the bottle and the reagent added. By this manner of addition, mechanical stirring at the sample surface is minimized. After the addition of reagents, the bottle is tightly stoppered and stored for analysis.

1 / To the samples collected in the BOD bottles, add 2 ml of the manganese sulfate solution, followed by 2 ml of the alkali iodide–azide reagent by the method described above. Stopper the BOD bottle and mix by inverting the bottle several times. A manganese hydroxide precipitate will form. Remix and allow to settle.

2 / When settling has produced 100 ml of clear supernate, add 2.0 ml of concentrated H_2SO_4 by the method described above. Restopper immediately and mix by inversion.

3 / The sample may now be removed from the BOD bottle for analysis. In calculating the volume of sample to be titrated, it is necessary to compensate for the loss of the original sample by addition of the reagents.

4 / Titrate 200 ml of sample (after correcting for addition of reagents) with 0.025 *N* thiosulfate solution to a pale straw color.

5 / After the straw color is obtained, add 2 ml of starch solution and continue the addition of thiosulfate until the first disappearance of the blue color. Note the volume of thiosulfate used.

Calculations / Since exactly 1 ml of 0.025 *N* thiosulfate is equivalent to 0.2 mg of DO, each milliliter of thiosulfate used in the titration is equivalent to 1 mg/liter of DO, assuming that a volume equivalent to 200 ml of original sample is titrated. Thus a direct reading of the buret will give mg/liter DO.

dissolved oxygen (alternative method for marine waters): method II

The following method is commonly used in the analysis of seawater samples. It is a modification of the previous method and also is based upon the titration of the sample with a sodium thiosulfate solution. In this method the sodium thiosulfate used in the titration is standardized and a correction factor, termed f, is obtained. The rationale of f corrections is similar to that discussed for α corrections (Method I).

Reagents

1 / Manganous sulfate reagent: dissolve 480 g of $MnSO_4 \cdot 4H_2O$ in distilled water and dilute to the mark in a 1-liter volumetric flask.

2 / Alkaline iodide solution: dissolve 500 g of NaOH in 500 ml of distilled water. Dissolve 300 g of potassium iodide in 450 ml of distilled water. Mix the two solutions.

3 / Starch-indicator solution: suspend 2 g of soluble starch in 400 ml of distilled water. Add, with vigorous stirring, a 20% NaOH solution until the starch solution becomes clear and allow to stand for 1 hour. Add concentrated HC1 until the solution is just acid (to litmus paper) and, at that time, add 2 ml of glacial acetic acid. Dilute this solution to 1 liter. This solution is to be discarded when the color (upon addition in the course of the titration) is no longer a pure blue. At this time it will take on a green or brownish tint.

4 / Standard thiosulfate solution (0.01 N): dissolve 2.9 g of $Na_2S_2O_3 \cdot 5H_2O$ and 1 g of Na_2CO_3 in distilled water, transfer to a 1-liter volumetric flask, and dilute to the mark. Add 1 drop of C_2S as a preservative. In this solution 1 ml of 0.01 N thiosulfate = 0.005 mg–at of O_2. If a large number of samples are to be analyzed, or samples will be analyzed periodically over a period of time, several liters of this solution may be prepared initially and stored in a well-stoppered dark bottle below 25°C.

5 / Iodate solution (0.0100 N): dissolve with gentle warming exactly 0.3567 g of KIO_3 in distilled water, cool, transfer to a 1-liter volumetric flask, and dilute to the mark. In this method the

standard sodium thiosulfate solution is "corrected" by titration against the iodate solution. The correction is in the form of a numerical f factor which must be used in all oxygen calculations.

f factor determination

1 / Fill a 300-ml BOD bottle (see Appendix III, Part B, Method A) with distilled water. Add 1.0 ml of concentrated H_2SO_4 and 1.0 ml of alkaline iodide solution; restopper and mix.

2 / Add 1.0 ml of manganous sulfate solution; restopper and mix.

3 / Transfer 50-ml aliquots (portions) and place in the titration flask.

4 / Add 5.00 ml of 0.0100 N iodate standard. Allow to stand 3 minutes. The temperature, at this time, must be below 25°C and out of direct sunlight.

5 / Titrate this solution with the 0.01 N thiosulfate solution.

6 / The f factor is calculated from the following equation in which v equals the volume of standard thiosulfate used to reach the end point:

$$f = \frac{5.00}{v}$$

Sampling and analytical procedure / It is necessary to use 300-ml BOD bottles in this method. These bottles should be rinsed twice, with portions of the sample being collected to prevent contamination. If the sample is obtained by use of a Nansen bottle (Chapter 10), a length of rubber tubing should be run from the top to the bottom of the BOD bottle. This will prevent mechanical stirring at the sample surface. Water is allowed to overflow from the top of the BOD bottle, which is then tightly stoppered. The water to be used for oxygen analysis should be the first samples drawn from the sample bottles.

The analysis should begin within 1 hour, since oxygen will be lost as the samples warm to room temperature, as well as by microbial respiration. The samples should be stored, if necessary, in the dark to minimize photosynthesis. If the samples must be held prior to analysis, they may be "pickled" by adding the manganous sulfate and alkaline iodide solutions.

It is recommended that the titration be carried out in 125-ml Erlenmeyer flasks. The outside bottoms and sides should be painted white. This will enable the analyst to recognize the end point more readily.

Procedure

1 / Remove the stopper from the BOD bottle and add 1.0 ml of manganous sulfate and 1.0 ml of alkaline iodide solution to the bottom of the sample (see p. 229). Restopper and mix until the precipitate is uniformly dispersed. Assure that there are no air bubbles in the bottle.

2 / The precipitate should settle in 2 to 3 minutes; at that time mix again. Allow the samples to stand until the precipitate has settled at least one third of the way down the bottle and the supernatent is clear.

3 / Add 1.0 ml of concentrated H_2SO_4, restopper, and mix until the precipitate is dissolved.

4 / Transfer 50.0 ml of the sample to the painted Erlenmeyer flasks. Titrate with the standard thiosulfate solution. Thiosulfate is added until a pale-straw-colored solution is obtained. Add 5 ml of the soluble starch solution. The solution will turn blue with this addition. Continue the titration until the blue disappears. Note the volume of thiosulfate used. The oxygen is calculated from the following formula, in which v equals the volume of thiosulfate added to reach the end point, and f is the correction factor obtained in the standardization of the thiosulfate:

$$\text{mg–at } O_2/\text{liter} = 0.1006 \times f \times v$$

NOTE—It is customary, in this method, to prepare a blank to compensate for any color that may be due to the reagents. If a slight coloration results, the blank is determined as follows.

1 / A BOD bottle is filled with distilled water and 1.0 ml of concentrated H_2SO_4 and 1.0 ml of alkaline iodide is added. The solution is then mixed.

2 / 1 ml of manganous sulfate is added and the solution remixed.

3 / 50.0 ml of this solution is placed in the Erlenmeyer flask.

4 / This solution is titrated with thiosulfate until colorless. The volume is noted. This volume serves as the blank and, if necessary, is subtracted from the final volume of the thiosulfate in both the sample titration and the standardization of the thiosulfate, prior to performing any further calculations (f factor and/or O_2 concentrations). If the blank exceeds 0.1 ml, the reagents should be discarded and new ones prepared.

part c: biological oxygen demand (standard methods)

The decomposition of sewage, sewage-plant effluents, and agricultural and industrial wastes will deplete the oxygen (increase the oxygen demand) in waters receiving these materials. The oxygen demand in any system is increased by the input of three broad classes of materials: (1) carbonaceous organic material, which is used as an energy source by aerobic organisms; (2) oxidizable nitrogen derived from nitrite, ammonia, and organic nitrogenous compounds, which serve as an energy source for specific microorganisms such as *Nitrosomonas* and *Nitrobacter* (see Chapters 2 and 4); and (3) chemical-reducing compounds such as Fe(III), sulfite, and sulfide, which react with and consume dissolved oxygen.

In raw and settled domestic sewage, the oxygen demand is brought about by the first class of materials, and the oxygen demand is biological in nature (BOD). In biologically treated effluents (see Chapter 9) the oxygen demand is also biological in nature and is brought about by the second class of materials.

Since most wastes are complex and contain a variety of organic compounds not readily decomposed by biological activity, the oxygen demand placed on a system to accomplish the complete decomposition of this material is not measured. Rather, a 5-day BOD analysis is performed, and this 5-day BOD is a generally accepted standard test. In many cases it is necessary to add microorganisms to the sample under investigation in order to obtain a valid BOD. This is termed *seeding*.

If seeding is necessary, the seed found to be satisfactory for the particular water under study is added to the dilution water (see 1 under Procedure). It is difficult to determine the seed to use without past experience. However, the following generalizations can be made: In many cases (food processing wastes, in particular) a satisfactory seed is obtained by using the supernatent from domestic sewage which has been stored at 20°C for 24 hours. Industrial wastes may not be affected by domestic sewage seed. In these cases receiving water collected 2 to 5 miles below the point of discharge may be used as a seed. Receiving water used as seed generally gives the best estimate of the effect of waste input on the oxygen levels. If receiving water is used as a seed, it must be collected at a point downstream that has built up populations capable of utilizing the various components present in the effluent.

Apparatus

1 / 300-ml BOD bottles. The same precautions should be taken in using these bottles for the BOD analysis as were exercised in the dissolved oxygen analysis.

2 / Incubator (either air or water): the incubator should be thermostatically controlled at $20°C \pm 1°C$. All light should be excluded to prevent photosynthesis.

Reagents

1 / Phosphate buffer solution: dissolve 8.5 g of KH_2PO_4, 21.75 g of K_2HPO_4, 33.4 g of $Na_2HPO_4 \cdot 7H_2O$, and 1.7 g of NH_4Cl in approximately 500 ml of distilled water, transfer to a 1-liter volumetric flask, and dilute to the mark. The pH of this solution should be 7.2 without further adjustment.

2 / Magnesium sulfate solution: dissolve 22.5 g of $MgSO_4 \cdot 7H_2O$ in distilled water, transfer to a 1-liter volumetric flask, and dilute to the mark.

3 / Calcium chloride solution: dissolve 27.5 g of anhydrous $CaCl_2$ in distilled water, transfer to a 1-liter volumetric flask, and dilute to the mark.

4 / Ferric chloride solution: dissolve 0.25 g of $FeCl_3 \cdot 6H_2O$ in distilled water, transfer to a 1-liter volumetric flask, and dilute to the mark.

5 / 1 liter of 1 N H_2SO_4.

6 / 1 liter of 1 N NaOH.

7 / Sodium sulfite solution (0.025 N): dissolve 1.575 g of anhydrous Na_2SO_3 in distilled water, transfer to a 1-liter volumetric flask, and dilute to the mark. This solution is unstable and should be prepared daily.

Procedure / The pH of the samples should be adjusted to 7 with 1 N H_2SO_4 or 1 N NaOH prior to beginning the analysis.

1 / Dilution water: the distilled water used as dilution water should be stored in cotton-plugged bottles for a sufficient length of time to have become saturated with dissolved oxygen. If necessary, the water may be aerated by bubbling air into the water.

2 / If necessary, the dilution water should be seeded at this point.

3 / Dilution of samples: It is necessary to make several dilutions of the sample to obtain the required oxygen depletions. The following dilutions are recommended: 0.1 to 1.0% for strong trade wastes, 1 to 5% for raw and settled sewage, 5 to 25% for oxidized effluents, 25 to 100% for polluted river waters.

a / Carefully siphon the required volume of dilution water (seeded if necessary) into a 1- or 2-liter graduated cylinder, avoiding the aeration of the water.

b / Add a sufficient quantity of sample to prepare the proper dilution (see 3 above). Mix well with a plunger-type mixing rod to avoid the addition of air.

c / Siphon the diluted sample into two BOD bottles.

d / One bottle is incubated at 20°C for 5 days.

e / The dissolved oxygen in the second bottle is determined immediately by the method given in Appendix III, Part B, or by the use of an oxygen meter. This will give the initial dissolved oxygen concentration of the sample.

f / Prepare succeeding dilutions of lower concentrations in the same manner. Using a raw sewage sample as an example: the initial dilution would prepare a 5% solution of raw sewage, while the succeeding dilutions would be performed to prepare 4.3, 2, and 1% solutions.

4 / All samples and blank dilution water samples (dilution water with no sample added) are incubated for 5 days at 20°C.

5 / After incubation, the oxygen levels of all samples and the blank dilution water is determined by the method given in Appendix III, Part B, or by use of an oxygen meter.

6 / Those dilutions showing a residual DO of at least 1 mg/liter and a depletion of at least 2 mg/liter are considered the most reliable.

Seed correction / If it is necessary to seed the sample to obtain a suitable oxygen reduction, the oxygen depletion of the seed may be determined as follows. A separate series of seed dilutions are set up (see 3 under Procedure above) and incubated. The samples resulting in a 40 to 70% reduction in oxygen are selected, and one of these samples is used to calculate the seed correction. The correction is calculated by the formula

$$\text{seed correction} = (B_1 - B_2) = f$$

The f factor is determined by the formula

$$f = \frac{\%\ \text{seed in}\ D_1}{\%\ \text{seed in}\ B_2}$$

in which D_1 = dissolved oxygen 15 minutes after preparation

B_1 = dissolved oxygen of seed before incubation

B_2 = dissolved oxygen of seed after incubation

f = ratio of seed in sample to seed in control

If applicable, the seed correction is then used in the final oxygen calculation.

part d: chemical oxygen demand (standard methods)

The determination of the chemical oxygen demand of a system is dependent upon the determination of those compounds that are capable of undergoing chemical oxidation. The decomposition of these materials is chemical in nature and serves to reduce the available oxygen within a system. The COD of a system varies with the water composition, temperature, and contact time. In some cases there is a degree of correlation between BOD and COD. Generally, however, chemical decomposition and biological decomposition are separate and distinct processes, and the results will vary.

Most types of organic matter are destroyed by boiling in a mixture of chromic acid and sulfuric acid. The amount of organic matter liberated is then determined by titration with ferrous ammonium sulfate and the COD calculated. The procedure follows:

Apparatus

1 / Reflux condensers.

2 / 500-ml round-bottomed flasks with ground-glass necks.

3 / Heating mantles.

Reagents

1 / Standard potassium dichromate solution (0.250 *N*): dissolve 12.259 g of $K_2Cr_2O_7$ (primary standard grade) in distilled water, transfer to a 1-liter volumetric flask, and dilute to the mark. Add 0.12 g of sulfamic acid to the standard to eliminate nitrate interference.

2 / Sulfuric acid reagent: dissolve 22 g of silver sulfate per 9-lb bottle H_2SO_4. It will take 1 to 2 days for the silver sulfate to dis-

solve. If this quantity of reagent is not required, the concentrations may be varied accordingly.

3 / Standard ferrous ammonium sulfate reagent (0.01 N): dissolve 39 g of analytical-grade $Fe(NH_4)_2 \cdot (SO_4)_2 \cdot 6H_2O$ in distilled water, add 20 ml of concentrated H_2SO_4, cool, transfer to a 1-liter volumetric flask, and dilute to the mark. This solution must be restandardized (see below) daily.

4 / Ferroin indicator solution: dissolve 1.485 g of 1, 10-phenanthroline monohydrate and 0.695 g of $FeSO_4 \cdot 7H_2O$ in distilled water and dilute to 100 ml.

5 / Mercuric sulfate crystals.

Standardization

1 / Dilute 10.0 ml of the standard potassium dichromate to 100 ml and add 30 ml of concentrated H_2SO_4. Allow to cool.

2 / Titrate with the ferrous ammonium sulfate titrant using 2 drops of ferrion as the indicator. The normality of the ferrous ammonium sulfate is determined by the formula

$$N = \frac{\text{ml of } K_2Cr_2O_7 \times 0.25\, N}{\text{ml of } Fe(NH_4)_2\,(SO_4)_2} \quad \text{(used to reach the end point)}$$

Procedure

1 / Place 0.4 g of $HgSO_4$ in the reflux flask (500-ml round-bottomed flask) and add 20.0 ml of the water sample. Swirl to mix.

2 / Add 10.0 ml of the standard potassium dichromate solution.

3 / Add 30 ml of sulfuric acid reagent.

4 / Add 5 to 6 boiling chips to prevent "bumping" during the refluxing.

5 / Attach the flask to the reflux condenser and reflux for 2 hours.

6 / After refluxing, the solution is cooled and the reflux condenser is washed down with a minimum of distilled water.

7 / Detach the flask from the condenser and measure the volume of the solution. Bring up to a volume of 140 ml with distilled water if necessary.

8 / Add 3 drops of the ferrion indicator and titrate to a reddish-brown end point with the standardized ferrous ammonium sulfate titrant.

9 / A blank consisting of 20 ml of distilled water in place of the sample is treated, refluxed, and titrated in the same manner.

Calculations / The COD is calculated from the equation

$$\text{mg/liter of COD} = \frac{(a - b)\, c \times 8000}{\text{ml of sample}}$$

in which a = ml of $Fe(NH_4)_2\,(SO_4)_2$ used for blank

b = ml of $Fe(NH_4)_2\,(SO_4)_2$ used for sample

c = ml of $Fe(NH_4)_2\,(SO_4)_2$ (determined in standardization with $K_2Cr_2O_7$)

part e: spectrophotometric methods of analysis

A number of spectrophotometric methods are available for the analysis of water and wastewater samples. The majority employed the use of either the visible spectrophotometer (generally the Spectronic 20) or the atomic-absorption spectrophotometer. In some cases, however, particularly when analyzing oceanic samples of low concentration, a spectrophotometer with a 100-mm light path may be required. This section deals only with commonly used methods that can be carried out with the minimum of time and equipment. The methods given here are for the commonly measured parameters of orthophosphate, nitrite, nitrate, detergents, and a brief section is included on metal analysis. By performing these analyses in conjunction with those methods given in Parts A through D, it is believed that accurate and valid determinations of water quality can be obtained on the majority of systems likely tó be encountered.

For additional methods, the following sources should be consulted: *Standard Methods for the Examination of Water and Wastewater* published by the American Public Health Association, New York; *Methods of Water Analysis,* published by the Environmental Protection Agency, Washington, D.C.; and *A Practical Handbook of Seawater Analysis* by Strickland and Parsons, published by the Fisheries Research Board of Canada, Ottawa.

nitrate–nitrite (environmental protection agency)

In this method the nitrate must be converted to nitrite prior to analysis. This conversion is accomplished by passing the water sample through a column packed with granulated copper–cadmium. This process converts the NO_3 to NO_2, which is then reacted with a chromogenic reagent and analyzed spectrophotometrically.

This method can be used to determine only the nitrite concentration (by analyzing the sample directly without passing it through the Cu–Cd column) or to determine both nitrate and nitrite. This is accomplished by analyzing a portion (A) of the sample directly as discussed above to determine the NO_2 concentration. The remainder of the sample (B) is then passed through the column, the NO_3 is converted to NO_2; the sample is analyzed and the NO_2 concentration in sample A is subtracted from the NO_2 concentration in sample B (NO_2 plus converted NO_3). This will give the NO_3 concentration (NO_3 converted to NO_2 in the column). The EPA recommends this method for the analysis of drinking water, groundwater, surface fresh water, marine waters, and domestic and industrial wastewater.

The analysis should be performed as soon as possible after the samples are collected to prevent any changes in nitrite and/or nitrate concentration due to biological activity. If the samples are to be stored prior to analysis, they may be preserved by refrigeration at 4°C for up to 24 hours. When they are stored for more than 24 hours, they should be preserved with H_2SO_4 (2 ml/liter of sample) and refrigerated.

Since nitrate and nitrite are dissolved in the water, the samples are commonly filtered (glass-fiber filter or a 0.45–μm membrane filter) to remove suspended matter. This is desirable, since the suspended material would tend to restrict the sample flow through the column if it were not removed. It is recommended that highly turbid samples be pretreated with zinc sulfate prior to filtration, to remove the majority of the particulate matter in the sample and to facilitate filtration. It is known that metals such as iron and copper will interfere with the analysis and give erroneous (low) results. In water samples known or suspected of containing these metals, the sample should be pretreated with EDTA to remove the metals.

It is to be noted that the EPA method determines both nitrate and nitrite. If only nitrate analysis is required, a separate determination (as discussed above) must be made for nitrite and the appropriate corrections made. The nitrite may be determined without passing the sample through the column (see Analysis of Samples, below).

The apparatus required for this analysis is a spectrophotometer for use at a monochrometer setting of 540 nm with a light path of 1 cm or longer and the Cd–Cu column. The column may be constructed of a 22-cm-long glass tube with a 3.5-mm inside diameter. The copperized cadmium (see below for preparation) is added to the column in sufficient quantity so that it fills at least 18.5 cm of its length. The reservoir (to contain the water and reagents) is attached to the upper portion and may be constructed of a 100-ml plastic sample bottle by merely cutting off the bottom portion. A section of Tygon tubing with a pinch or screw clamp provides a convenient stopcock on the opposite end. It is recommended that a screw clamp be used in order to control the flow rate more efficiently. A 100-ml pipet can be used in place of the glass tubing if so desired. The apparatus is depicted in Fig. I–1.

Preparation of copperized cadmium (column packing)

1 / Granulated cadmium: 40–60 mesh

2 / Copper–cadmium: the granulated Cd is cleaned with 6 *N* HCl and copperized with a 2% by weight solution of $CuSO_4$ in the following manner:

a / Wash the Cd with 6 *N* HCl and rinse with distilled water.

b / Swirl 25-g portions of Cd in 100-ml portions of 2% $CuSO_4$ until the initial blue color fades. Decant and repeat with fresh $CuSO_4$.

c / Wash the Cu–Cd with distilled water (at least 10 washings are required) to remove all the precipitated copper.

Column preparation

1 / Insert a glass-wool plug into the bottom of the column and fill with distilled water. Add sufficient Cu–Cd to produce a column 18.5 cm in length. It is necessary to maintain a level of distilled water above the Cu–Cd to prevent the entrapment of air.

2 / Pass 200 ml of *dilute* ammonium chloride–EDTA solution.

a / The stock solution is prepared as follows: dissolve 13 g of ammonium chloride and 1.7 g of disodium ethylene diamine-tetraacetatic acid in 900 ml of distilled water. Adjust pH to 8.5 with concentrated ammonium hydroxide, transfer to a 1-liter volumetric flask, and dilute to the mark.

b / The dilute solution is prepared by diluting 300 ml of the stock solution to 500 ml with distilled water in a volumetric flask.

3 / After passing the dilute solution of ammonium chloride–EDTA through the column, the column is activated by passing 100 ml of a solution composed of 25 ml of a 1.0 mg/liter NO_3–N standard mixed with 75 ml of the *stock* ammonium chloride–EDTA solution through the column at a flow rate of 7 to 10 ml/minute. After this has been accomplished, the column is activated.

Reagents / The following reagents are necessary for the analysis:

1 / Color reagent: dissolve 10 g of sulfanilimide and 1 g of *N*(1-naphthyl)-ethylenediamine dihydrochloride in a prepared solution of 100 ml of concentrated phosphoric acid in 800 ml of distilled water and dilute to 1 liter with distilled water.

2 / Zinc sulfate solution: dissolve 100 g of $ZnSO_4 \cdot 7H_2O$ in distilled water, transfer to a 1-liter volumetric flask, and dilute to the mark.

3 / 6 *N* NaOH.

4 / Concentrated ammonium hydroxide.

5 / Stock nitrate solution: dissolve 7.218 g of KNO_3 in distilled water, transfer to a 1-liter volumetric flask, and dilute to the mark. Preserve with 2 ml of chloroform per liter. In this solution 1.0 ml = 1.00 mg of NO_3–N. This solution is stable for 6 months.

6 / Standard nitrate solution: dilute 10.00 ml of the stock nitrate solution to 1 liter with distilled water. In this solution 1.0 ml = 0.01 mg of NO_3–N.

7 / Stock nitrite solution: dissolve 6.072 KNO_2 in distilled water, transfer to a 1-liter volumetric flask, and dilute to the mark. Preserve with 2 ml of chloroform and refrigerate. In this solution 1.0 ml = 1.00 mg of NO_2–N. This solution is stable for approximately 3 months.

8 / Standard nitrite solution: dilute 10.0 ml of stock nitrite solution to 1 liter with distilled water. In this solution 1.0 ml = 0.01 mg of NO_2–N.

Preparation of standards / Using the standard nitrate solution, prepare the following in 100-ml volumetrics.

ml of nitrate standard/100 ml of solution	*Conc. (mg of* No_3*–N/liter)*
0.00	0.0
0.05	0.5
0.10	1.0
0.20	2.0
0.50	5.0
1.00	10.0

Standardization of the spectrophotometer

1 / Set monochrometer to 540 nm.

2 / Blank and zero the instrument.

3 / Add 75.0 ml of stock ammonium chloride–EDTA solution to 25.0 ml of each standard mix.

4 / Pour the sample into the activated column; the flow rate should be 7 to 10 ml/min.

5 / Discard the first 25 ml to avoid cross-contamination. Collect the remainder.

6 / Add 2.0 ml of the sulfanilimide; *N*(1-napthyl)-ethylenediamine dihydrochloride color reagent to the collected standard.

7 / Add 10 minutes for color development. After the color is developed, the sample must be analyzed on the spectrophotometer.

8 / Record the absorbance versus concentration of each standard and plot this data to prepare the standard nitrate curve.

9 / If nitrite data are desired, prepare a similar curve using diluted nitrite standard. In this case, however, it is not necessary to run the nitrite through the column prior to carrying out the chromogenic reaction and spectrophotometric analysis.

Analysis of samples / The analysis of samples is carried out exactly as the procedure described for the standards. In other words, a 25-ml sample is added to 75 ml of ammonium chloride–EDTA solution and run through the column. The sample is then reacted with the color reagent and analyzed, and the concentration is determined from the standard curve. In these cases, however, it is advisable to blank and zero the instrument and then place a portion of the sample in a cuvette with no reagents added. The absorbance given by this sample, if significant, should then be subtracted from the absorbance given by the samples reacted with the chromogenic reagent. This will compensate for any background color due to the sample.

Turbidity removal / Either of the following procedures may be used.

1 / Filter the sample through a glass-fiber filter or a 0.45-μm membrane filter. It is good practice to routinely filter all samples upon collection.

2 / Add 1 ml of the zinc sulfate solution (number 2 under reagents) to 100 ml of sample and mix. Add 0.5 ml of 6 *N* NaOH to obtain a pH of 10.5. Allow the sample to stand 5 minutes and filter.

Oil and grease removal / Filter 100 ml of the sample and adjust the pH to 2 by addition of concentrated HCl. Add the sample to a separatory funnel, add 25 ml of chloroform, allow the sample to layer, and draw off the chloroform/oil fraction. Repeat this extraction twice.

pH adjustment / If the pH is below 5 or above 9, it is necessary to adjust the pH to between 5 and 9 with either concentrated HCl or concentrated H_2SO_4; the pH may also be adjusted by using a base such as concentrated NH_4OH. Adjustment of pH must be accomplished prior to the addition of the ammonium chloride–EDTA solution.

orthophosphate (modified USGS method)

This method may be used for the analysis of drinking water, groundwater, surface and marine waters, and domestic and industrial wastewater. The orthophosphate is reacted with an appropriate chromogenic reagent and analyzed spectrophotometrically. In the case of deep-ocean-water analysis, where the phosphate is present in low concentrations, a light path longer than that available with the common Spec 20 may be necessary. In all other samples the common 1-mm light path should suffice.

Preparation of reagents and standards

1 / Ammonium molybdate solution: dissolve 15 g of ammonium molybdate $(NH_4)_6\ MO_7 \cdot 4H_2O$ in distilled water; transfer to a 500-ml volumetric flask and dilute to the mark. This solution is stable indefinitely.

2 / Sulfuric acid solution: add 140 ml of concentrated H_2SO_4 to 900 ml of distilled water. Allow to cool and store in a glass bottle.

3 / Ascorbic acid solution: dissolve 27 g of ascorbic acid in 500 ml of distilled water. This solution is stored frozen in a plastic reagent bottle. Thaw for use in the analysis and refreeze immediately. This solution is stable for several months but should not be kept unfrozen for any appreciable time.

4 / Potassium antimonyl–tartrate solution: dissolve 0.34 g of potassium antimonyl tartrate in distilled water, transfer to a 250-ml volumetric flask, and dilute to the mark. This solution may be warmed to place all the solute in solution. The solution is stable for several months.

5 / Mixed reagent: this solution is prepared immediately prior to the actual analysis. Mix 100 ml of ammonium molybdate, 250 ml of sulfuric acid, and 50 ml of the potassium antimonyl tartrate solutions. This quantity is suitable for the analysis of 50 samples and should be adjusted as necessary.

6 / Stock phosphate solution 1: dissolve 4.986 g of anhydrous KH_2PO_4 in distilled water, transfer to a 1-liter volumetric flask, and dilute to the mark. Add 1 ml of chloroform as a preservative and store in a dark bottle. In this solution 1.00 ml = 36 μg-at PO_4.

7 / Stock phosphate solution 2: dilute 10.0 ml of this solution to 1 liter, preserve with 1 ml of chloroform, and store in a dark bottle. In this solution 3.6 μg-at PO_4/liter (in a 100-ml sample).

8 / To prepare the top standard (most highly concentrated standard), dilute 50 ml of stock phosphate solution 2 to 1 liter with distilled water. This solution is equal to 18.0 μg-at PO_4/liter.

9 / By a series of dilutions from this solution, prepare 100 ml of the following standards: 13.5 μg-at PO_4/liter; 9 μg-at PO_4/liter; 4.5 μg-at PO_4/liter; 2.25 μg-at PO_4/ liter; 1.12 μg-at PO_4/liter, and 0.56 μg-at PO_4/liter.

NOTE—In the analysis of samples containing high PO_4 concentrations, the standards of low concentrations may be omitted. In the analysis of samples containing low concentrations of PO_4, the upper standards may be omitted. In field work involving the analysis of a large number of diverse samples, however, it is generally beneficial to prepare a standard curve encompassing all the standards.

Standardization of the spectrophotometer

1 / Set the monochrometer at 885 nm (in the case of the Spectronic 20, it is necessary to use a red filter and the appropriate phototube to work at this wavelength).

2 / Blank and zero the instrument.

3 / To 100 ml of each standard, add exactly 10.00 ml of mixed reagent and mix. Allow the color to develop for 5 minutes.

4 / Place in a cuvette and analyze on the spectrophotometer.

5 / Record the concentration and absorbance of each standard and plot these data to prepare the standard PO_4 curve.

Analysis of samples / A 100-ml portion of each sample is reacted with the mixed reagent and the absorbance determined. The concentration is then determined from the standard curve. It is generally also advisable to blank and zero the instrument and then determine the absorbance of each sample with no reagent added in order to compensate for any background color due to the sample (see NO_2–NO_3 method and Chapter 11 for details).

detergent analysis (standard methods): modified

The most widely used method of detergent analysis is the *methylene blue process.* This method depends on the formation of a blue-colored solution, when the methylene blue reacts with the synthetic detergents, commonly termed *syndets.* The syndet–methylene blue is soluble in chloroform ($CHCl_3$) and insoluble in water, whereas the methylene blue and syndet (in their uncombined form) are insoluble in $CHCl_3$ and soluble in water. Thus the syndet is combined with methylene blue, extracted from the water sample into the chloroform, and analyzed spectrophotometrically.

One problem that may be encountered in this analysis is that many naturally occurring substances also form chloroform-soluble substances with the methylene blue. Both organic and inorganic materials are known to react and give either positive and/or negative interferences. Therefore, when analyzing raw water samples, additional methods may have to be employed to remove these substances prior to initiating the analysis. See the AWWA Task Group Report 2662P for "clean-up" techniques.

Reagents

1 / Standard syndet stock solution: dissolve 1.00 g of syndet in distilled water, transfer to a 1-liter volumetric flask, and dilute to the mark. This solution contains 1 mg/ml.
   a / From this stock solution, by a series of dilutions, prepare 1 liter of the following standard solutions: 0.75 mg/ml, 0.50 mg/ml, 0.25 mg/ml, 0.125 mg/ml, 0.062 mg/ml, and 0.31 mg/ml.
   b / These standards are analyzed by the method given below, the absorbance determined, and the standard curve plotted.

2 / Aqueous phenolpthalein indicator.

3 / 1 *N* NaOH.

4 / 1 *N* H_2SO_4.

5 / Chloroform ($CHCl_3$).

6 / Methylene blue reagent: dissolve 0.1 g of methylene blue (Eastman P573) in 100 ml of distilled water. Transfer 30 ml of this solution to a 1-liter volumetric flask and add 500 ml of concentrated H_2SO_4 and 50 g of $NaH_2PO_4 \cdot H_2O$. Mix until dissolved and dilute to the mark with distilled water.

7 / Wash solution: place 6.8 ml of concentrated H_2SO_4 with 500 ml of distilled water in a 1-liter volumetric flask. Dissolve 50 g of $NaH_2PO_4 \cdot H_2O$ in distilled water and add this solution to the volumetric flask. Dilute to the mark with distilled water.

Procedure

1 / Place 100 ml of sample (or standard if the curve is to be prepared) in a beaker.

2 / Using phenolpthalein as the indicator, make the sample basic with 1 *N* NaOH.

3 / Acidify with 1 *N* H_2SO_4.

4 / Place each sample and/or standard in a separatory funnel; add 10 ml of $CHCl_3$ and 25 ml of methylene blue. Shake for 30 seconds; release the pressure and allow to separate.

5 / Draw off the $CHCl_3$ layer (bottom layer) into another separatory funnel and, in the first separatory funnel, repeat the extraction using 10 ml of $CHCl_3$, three times. When the extraction is repeated, $CHCl_3$ is added but not methylene blue. If the blue color in the aqueous phase disappears or becomes faint, however, 25 ml of methylene blue may be added.

6 / After each extraction the $CHCl_3$ layer is drawn off into the second separatory funnel.

7 / After all extractions have been performed, add 50 ml of the wash solution to the second funnel and shake for 30 seconds; allow to settle and draw off the chloroform layer, through glass wool, into a 100-ml volumetric flask.

8 / Repeat the washing twice with 10 ml of chloroform added each time and draw this material off into the 100-ml volumetric flask. Dilute to the mark with chloroform and mix.

9 / Determine the absorbance of each sample (standard) at 652 nm against a chloroform blank.

10 / The standards are plotted and a standard curve is prepared.

11 / The concentrations of the samples are then determined from the standard curve.

trace-metal analysis

Trace metals in water, tissue, and sediment samples are commonly determined by either visible spectrophotometry or by atomic-absorption spectrophotometry. Recently, the majority of these analyses have been carried out by atomic absorption spectrophotometry, because of its greater sensitivity and ease of analysis. The methods given here are for the atomic-absorption spectrophotometer. Visible spectrophotometric methods can be found in any textbook of instrumental analysis.

Water to be examined for the presence of trace metals can generally be analyzed directly. In some cases the metal may have to be concentrated into a smaller volume by extraction into an appropriate solvent prior to analysis. In either case, the atomic absorption is standardized by running a series of standards prior to the analysis of the samples. Appendix III, Part D, as well as the manufacturer's instructions, should be consulted prior to initiating the analysis.

Trace Metals (in sediments and tissues) / In the analysis of sediments and tissues, the metal must first be extracted from this material prior to analysis. The tissue or sediment is dried and the dry weight determined. The material is then refluxed in 100 ml of either 1 *N* HC or 1 *N* HNO_3 for 1 hour. This extraction is generally sufficient to remove all the acid-extractable metal and place it in solution.

A standard curve is prepared for the particular metals of interest. The concentration of the standards is generally given in mg/ml of solvent. After the refluxing has been completed, the supernatent is analyzed on the atomic-absorption spectrophotometer, and the data are obtained in mg/ml from the standard curve. Since the data must be equated to the original sample (sediment, tissue, sludge), the percentage of the particular metal by weight of the total sample must be calculated. This is accomplished by use of the following equation:

$$\frac{\text{mg/ml of metal}}{\text{total wt of sample}} \times 100 = \%\ \text{wt of metal}$$

It is to be noted that highly volatile metals, such as mercury, cannot be refluxed without danger of loss. In these cases the atomic-absorption spectrophotometer may be operated with an attachment that will enable the total sample to be analyzed without the reflux step.

In all cases the manufacturer's instructions should be consulted prior to performing the analysis. In addition, it is to be noted that the data can be determined in terms of ppm, ppb, mg/ml, or percent by weight, depending upon the units used in the preparation of the standard curve.

glossary

aerobic decomposition

The decomposition of materials by microorganisms in the presence of oxygen.

anaerobic decomposition

The decomposition of materials by microorganisms in the absence of oxygen.

aquifer

That portion of the subsoil which is permeable, porous, and yields water to wells.

aquitard

An impermeable subsurface barrier that prevents the vertical movement of groundwater.

atom

An electrically neutral particle composed of subatomic particles such as electrons, neutrons, and protons.

Avogadro's number

The number of particles (6.02×10^{23}) contained in 1 mole of a substance. When referring to molecules or atoms, it is also equivalent to the gram atomic or gram molecular weight expressed in grams.

bar

Mounds of sediment located offshore parallel to the berm.

benthic zone

The sediments that underlie a water column.

berm

The above-water portion of a beach.

BOD

Biological oxygen demand, the removal of oxygen from a system during decomposition.

chlorinity

The concentration of chloride ions found in 1 kg of seawater.

COD

Chemical oxygen demand, the removal of oxygen from a system due to chemical processes.

cone of depression

The depression in a water table in the vicinity of pumping operations. Caused by excessive removal of groundwater.

coriolis effect

The effect exerted by a rotating earth on the direction of water flow. Because of the Coriolis effect, water appears to travel to its right in the northern hemisphere and to its left in the southern hemisphere.

covalent bond

A chemical bond formed by the sharing of electrons.

Darcy's law

A geophysical law covering the relationship among permeability, groundwater pressure, and motion.

density

The relationship of mass to volume. Density equals mass per unit volume or $D = M/V$.

drawdown

The lowering of a water table due to the removal of groundwater.

dysphotic zone

The zone beneath sunlight penetration.

electron

A negatively charged subatomic particle found orbiting the nucleus of an atom.

epilimnion

The upper zone of gradual temperature change.

estuary

A semienclosed body of water with a free connection with the ocean.

euphotic zone

The upper zone of sunlight penetration.

euryhaline

A wide tolerance to salinity variation.

eurythermal

A wide tolerance to temperature variation.

eutrophic

A middle-aged system. Relatively shallow, warmer, and higher in nutrients than a young system.

facultative microorganism

A microorganism capable of carrying on its functions either in the presence or absence of oxygen.

food chain

The transfer of energy from plants to animals to microorganisms and then back to plants.

halocline

An abrupt change of salinity from the surface to the subsurface waters.

herbivore

An animal feeding exclusively on plants.

homothermous

Uniform temperature.

hydrogen bond

A weak intramolecular attractive force that results from the attractions exerted by the positive and negative regions of adjacent molecules.

hydrologic cycle

The process whereby liquid water or ice is converted to water vapor, transported in the vapor state, condensed, and returned to earth as precipitation.

hypersaline

A system of very high salinity.

hypolimnion

The lowest zone in a lake. This zone is characterized by constant low temperatures.

infiltration

The process whereby water sinks into the soil surface.

intertidal zone

That area alternately covered by water at high tide and exposed at low tide.

ion

A particle that has gained or lost electrons and has become charged.

ionic bond

A chemical bond formed through the complete gain and loss of electrons.

lentic

Standing-water systems, such as lakes and ponds.

longshore current

The current moving along shore in the surf zone.

lotic

Running-water systems, such as streams and rivers.

mole

The gram atomic or gram molecular weight of atoms or molecules expressed in grams. The mole is known to contain 6.02×10^{23} particles.

molecule

An electrically neutral substance composed of atoms chemically bonded together.

nonpoint source

The discharge of contaminants from poorly defined and scattered sources.

nonpolar molecule

A molecule in which there is neither complete gain and loss nor equal sharing of electrons. A molecule exhibiting complete electrical neutrality.

nutrient cycles

The cycling of essential materials such as phosphorus and nitrogen from plants to animals to decomposers, and then back into a suitable form for use by plants.

oligotrophic

A young system generally deep, cold, and low in nutrients.

overturn periods

Periods when a system becomes homothermous and capable of mixing.

permeability

The ability of sediments to transmit water.

photosynthesis

The conversion of carbon dioxide and water into carbohydrate and oxygen by green plants in the presence of sunlight.

phytoplankton

Plant plankton.

plankton

Organisms, either plant or animal, that are incapable of directed sustained motion within a water column. These organisms are moved about by the tides, currents, and the like.

point source

The input of contaminants from well-defined, easily identified sources.

polar molecule

A molecule in which there is an unequal sharing of electrons. This gives rise to positive and negative sites about the molecule.

pressure gradient current

The movement of water from an area of high pressure to an area of lower pressure.

protists

Microorganisms.

salinity

The total amount of dissolved material present in 1 kg of water.

senescent

An old system, generally very shallow, with water temperatures approximating atmospheric temperatures.

stagnation periods

Occur as a result of the water-temperature differences between the epilimnion and the hypolimnion. During these periods a temperature–density barrier prevents mixing of the water masses.

subsidence

The sinking of the land surface to fill the void created by groundwater removal.

surface runoff

Water flowing over the land surface and into streams.

suspended water

Water trapped by soil, rock, and so on, and held in the upper soil zone.

thermocline

The zone of rapid temperature change immediately below the epilimnion. In this zone the temperature changes by at least 1°C per meter depth.

transpiration

The conversion of liquid water to water vapor by plants.

water table

The upper limit of useable groundwater.

zone of aeration

The upper soil zone, where the spaces between soil particles are filled with a mixture of air and water.

zone of saturation

Generally located below the zone of aeration. In this zone the spaces between the soil particles are completely filled with water.

zooplankton

Animal plankton, such as larvae and jellyfish.

index